Application of Novel Methods for Mycotoxins Analysis

Application of Novel Methods for Mycotoxins Analysis

Editors

Veronica Maria Teresa Lattanzio
Biancamaria Ciasca

MDPI • Basel • Beijing • Wuhan • Barcelona • Belgrade • Manchester • Tokyo • Cluj • Tianjin

Editors

Veronica Maria Teresa Lattanzio
National Research Council of Italy - Institute of Sciences of Food Productionn
Italy

Biancamaria Ciasca
National Research Council of Italy - Institute of Sciences of Food Production
Italy

Editorial Office

MDPI
St. Alban-Anlage 66
4052 Basel, Switzerland

This is a reprint of articles from the Special Issue published online in the open access journal *Toxins* (ISSN 2072-6651) (available at: https://www.mdpi.com/journal/toxins/special_issues/mycotoxins_analysis).

For citation purposes, cite each article independently as indicated on the article page online and as indicated below:

LastName, A.A.; LastName, B.B.; LastName, C.C. Article Title. *Journal Name* **Year**, *Volume Number*, Page Range.

ISBN 978-3-0365-3523-4 (Hbk)
ISBN 978-3-0365-3524-1 (PDF)

Contents

About the Editors

Veronica Maria Teresa Lattanzio (Dr.). Chemist. Senior Researcher at the Institute of Sciences of Food Production—National Research Council of Italy. Coordinator of FOODSAFETY4EU: Multi-Stakeholder Platform for Food Safety in Europe (H2020, G.A. 101000613). As a food safety expert, her main research topic is the development and validation of analytical methods for mycotoxin and pesticide detection based on mass spectrometry techniques and immunoassays, including the organization of collaborative trials. She is a member of the Working Group "Biotoxins" (CEN/TC 275 WG5 "Food Analysis—Horizontal Methods–Biotoxins) of the European Committee for Standardization. She collaborates with the International Atomic Energy Agency (IAEA) as an expert to provide individual and collective training related to testing chemical contamination for food safety.

Biancamaria Ciasca (Dr.) Chemist. Researcher at the Institute of Sciences of Food Production—National Research Council of Italy. Her main research topic is the development and standardization of methods for the analysis of mycotoxins and pesticides in food/feed matrices based on advanced mass spectrometry techniques and immunoassays, including the preparation and characterization of reference materials. She is involved in the development of targeted and untargeted approaches for plant metabolomic studies as well as the integration of open-source tools for data processing and interpretation. She has expertise in social lab organization and facilitation.

 toxins

MDPI

Editorial

Introduction to This Special Issue of *Toxins*: Application of Novel Methods for Mycotoxin Analysis

Veronica M. T. Lattanzio * and Biancamaria Ciasca

Institute of Sciences of Food Production, National Research Council of Italy, Via Amendola, 122/O, 70126 Bari, Italy; biancamaria.ciasca@ispa.cnr.it
* Correspondence: veronica.lattanzio@ispa.cnr.it

Citation: Lattanzio, V.M.T.; Ciasca, B. Introduction to This Special Issue of *Toxins*: Application of Novel Methods for Mycotoxin Analysis. *Toxins* **2022**, *14*, 190. https://doi.org/10.3390/toxins14030190

Received: 23 February 2022
Accepted: 2 March 2022
Published: 4 March 2022

Crop contamination by mycotoxins is a global problem that poses significant economic burdens due to the food/feed losses that are caused by reduced production rates; the resulting adverse effects on human and animal health and productivity; and the trade losses associated with the costs incurred by inspection, sampling, and analysis before and after shipments. In this scenario, the development of fit-for-purpose analytical methods for regulated and (re)-emerging mycotoxins continues to be a dynamic research area. Some of the current trends in this research area are presented in the papers that have been selected for this Special Issue of *Toxins*.

The collected contributions address either the need for improved methods for mycotoxin detection addressed by new or incoming regulation (ergot alkaloids and *Alternaria* toxins) as well as methods for the detection of multiple mycotoxins. New approaches to enhance the performance of well-established methodologies, such as the enzyme-linked immunosorbent assay (ELISA) and fluorescence polarization immunoassays (FPIA), have already been proposed.

The recently issued European Commission Regulation (EU) 2021/1399, which sets the maximum limits for the sum of 12 main ergot alkalois (EAs), has been a driver for the development of improved analytical approaches for monitoring and official control purposes. Analytical challenges related to EA detection have been discussed by Lattanzio et al. [1], who reported on EA monitoring data in cereal and cereal-derived products collected in Italy over the period of 2017–2020 for official control purposes. To this scope, the authors set up and applied a method upon the verification of its fitness for in-house validation purposes. Poapolathep et al. [2] explored the liquid chromatography tandem mass spectrometry (LC-MS/MS) performance and applicability of EA detection in swine and dairy feeds, revealing a significant number of contaminated samples, regardless of whether the contamination was at EU regulation-compliant levels. Indeed, this new regulation also calls for methods that can be implemented for quick compliance testing. An interesting approach for EA screening was proposed by Kuner et al. [3], whose proposed method is based on EA cleavage by hydrazinolysis to convert them in a lysergic acid derivative, allowing their total content (sum of 12 EA) in food and feed to be quantified.

Though not yet regulated, *Alternaria* toxins (ATs) have been included in European Commission (EC) monitoring programmes since 2012. To fulfil this requirement and to complement the available LC-MS-based methods, research made steps toward the development of AT antibodies. Addante-Moya et al. [4] prepared and characterized two rationally designed synthetic haptens of *Alternaria* mycotoxins, which led to high-affinity antibodies of alternariol and alternariol monomethyl ether. These findings will pave the way for new immunoassay developments.

FPIA is a widely used homogeneous-based immunoassay with simple and rapid operational procedures. With the goal of keeping the sample preparation procedure as simple as possible, achieving improved sensitivity and selectivity is all about choosing the best tracer/antibody combination. In the study by Huang et al. [5], different antibody/tracer combinations were tested to set up a FPIA for the detection of ochratoxin A in rice. The resulting immunoassay fulfilled the mycotoxin screening and testing method validation and performance criteria set by the EU.

Evaluating contaminated food directly for specific fungi via the genes involved in aflatoxin production is a promising strategy. Based on this alternative approach, Elsayed Hafez et al. [6] developed a recombinant AflR gene (involved in aflatoxin biosyntesis) antiserum ELISA for the detection of aflatoxin-producing fungi in contaminated food.

Standardization is a challenging journey, especially when validating multi-mycotoxin methods. De Girolamo et al. [7] reported on this process within the M/520 standardization mandate of the European Commission. An LC-MS/MS method for the simultaneous determination of trichothecenes and zearalenone in wheat, wheat flour, and wheat crackers was validated through a collaborative study involving 15 participants from 10 countries. The results proved that the candidate method was fit for enforcement purposes.

Overall, even though not comprehensive, the collected manuscripts provide an up-to-date picture of the current trends in novel mycotoxin analysis methods, with LC-MS continuing to be the technique of choice for multi-mycotoxin detection, and testing new immunoreagents (labels and/or antibodies) represents a key step for improving the performance of screening methods.

Funding: This research received no external funding.

Acknowledgments: The editors wish to acknowledge all of the authors who contributed to this Special Issue and the expert peer reviewers for performing careful and rigorous evaluations.

Conflicts of Interest: The authors declare no conflict of interest.

References

1. Lattanzio, V.M.T.; Verdini, E.; Sdogati, S.; Caporali, A.; Ciasca, B.; Pecorelli, I. Undertaking a New Regulatory Challenge: Monitoring of Ergot Alkaloids in Italian Food Commodities. *Toxins* **2021**, *13*, 871. [CrossRef] [PubMed]
2. Poapolathep, S.; Klangkaew, N.; Zhang, Z.; Giorgi, M.; Logrieco, A.F.; Poapolathep, A. Simultaneous Determination of Ergot Alkaloids in Swine and Dairy Feeds Using Ultra High-Performance Liquid Chromatography-Tandem Mass Spectrometry. *Toxins* **2021**, *13*, 724. [CrossRef] [PubMed]
3. Kuner, M.; Kühn, S.; Haase, H.; Meyer, K.; Koch, M. Cleaving Ergot Alkaloids by Hydrazinolysis—A Promising Approach for a Sum Parameter Screening Method. *Toxins* **2021**, *13*, 342. [CrossRef] [PubMed]
4. Addante-Moya, L.G.; Abad-Somovilla, A.; Abad-Fuentes, A.; Agulló, C.; Mercader, J.V. Assessment of the Optimum Linker Tethering Site of Alternariol Haptens for Antibody Generation and Immunoassay Development. *Toxins* **2021**, *13*, 883. [CrossRef] [PubMed]
5. Huang, X.; Tang, X.; Jallow, A.; Qi, X.; Zhang, W.; Jiang, J.; Li, H.; Zhang, Q.; Li, P. Development of an Ultrasensitive and Rapid Fluorescence Polarization Immunoassay for Ochratoxin A in Rice. *Toxins* **2020**, *12*, 682. [CrossRef] [PubMed]
6. Hafez, E.; El-Aziz, N.M.A.; Darwish, A.M.G.; Shehata, M.G.; Ibrahim, A.A.; Elframawy, A.M.; Badr, A.N. Validation of New ELISA Technique for Detection of Aflatoxin B1 Contamination in Food Products versus HPLC and VICAM. *Toxins* **2021**, *13*, 747. [CrossRef] [PubMed]
7. De Girolamo, A.; Ciasca, B.; Pascale, M.; Lattanzio, V.M.T. Determination of Zearalenone and Trichothecenes, Including Deoxynivalenol and Its Acetylated Derivatives, Nivalenol, T-2 and HT-2 Toxins, in Wheat and Wheat Products by LC-MS/MS: A Collaborative Study. *Toxins* **2020**, *12*, 786. [CrossRef] [PubMed]

Article

Development of an Ultrasensitive and Rapid Fluorescence Polarization Immunoassay for Ochratoxin A in Rice

Xiaorong Huang [1,2,3,4,5,†], Xiaoqian Tang [1,2,3,4,5,†], Abdoulie Jallow [1,2,3], Xin Qi [5], Wen Zhang [5], Jun Jiang [5], Hui Li [2,5], Qi Zhang [1,2,3,4,5,*] and Peiwu Li [1,2,3,4,5,*]

1. Oil Crops Research Institute, Chinese Academy of Agricultural Sciences, Wuhan 430062, China; huangxiaorong@caas.cn (X.H.); tangxiaoqian@caas.cn (X.T.); 2019y90200006@caas.cn (A.J.)
2. Key Laboratory of Biology and Genetic Improvement of Oil Crops, Ministry of Agriculture, Wuhan 430062, China; lihui04@caas.cn
3. Laboratory of Quality & Safety Risk Assessment for Oilseed Products (Wuhan), Ministry of Agriculture, Wuhan 430062, China
4. Key Laboratory of Detection for Mycotoxins, Ministry of Agriculture, Wuhan 430062, China
5. Quality Inspection & Test Center for Oilseed Products, Ministry of Agriculture, Wuhan 430062, China; qixin@caas.cn (X.Q.); zhangwen@oilcrops.cn (W.Z.); jiangjun@caas.cn (J.J.)

* Correspondence: zhangqi01@caas.cn (Q.Z.); peiwuli@oilcrops.cn (P.L.); Tel.: +86-27-8681-2943 (P.L.)

† These authors contributed equally to this work.

Received: 28 August 2020; Accepted: 18 October 2020; Published: 29 October 2020

Abstract: Ochratoxin A (OTA) is a known food contaminant that affects a wide range of food and agricultural products. The presence of this fungal metabolite in foods poses a threat to human health. Therefore, various detection and quantification methods have been developed to determine its presence in foods. Herein, we describe a rapid and ultrasensitive tracer-based fluorescence polarization immunoassay (FPIA) for the detection of OTA in rice samples. Four fluorescent tracers OTA-fluorescein thiocarbamoyl ethylenediamine (EDF), OTA-fluorescein thiocarbamoyl butane diamine (BDF), OTA-amino-methyl fluorescein (AMF), and OTA-fluorescein thiocarbamoyl hexame (HDF) with fluorescence polarization values (δFP = FPbind-FPfree) of 5, 100, 207, and 80 mP, respectively, were synthesized. The tracer with the highest δFP value (OTA-AMF) was selected and further optimized for the development of an ultrasensitive FPIA with a detection range of 0.03–0.78 ng/mL. A mean recovery of 70.0% to 110.0% was obtained from spiked rice samples with a relative standard deviation of equal to or less than 20%. Good correlations (r^2 = 0.9966) were observed between OTA levels in contaminated rice samples obtained by the FPIA method and high-performance liquid chromatography (HPLC) as a reference method. The rapidity of the method was confirmed by analyzing ten rice samples that were analyzed within 25 min, on average. The sensitivity, accuracy, and rapidity of the method show that it is suitable for screening and quantification of OTA in food samples without the cumbersome pre-analytical steps required in other mycotoxin detection methods.

Keywords: FPIA; mycotoxin; OTA; detection methods; food safety; monoclonal antibody (mAb); tracer; HPLC

Key Contribution: Herein, we synthesized an OTA-AMF fluorescent tracer with 207 δFP value, and established an ultrasensitive fluorescence polarization immunoassay with a detection range of 0.03 to 0.78 ng/mL, providing an ultrasensitive, simple, and rapid detection method for on-site monitoring of OTA.

1. Introduction

Ochratoxin A (OTA) is the poisonous secondary metabolite excreted by *Penicillium* and *Aspergillus* species, which is often found in a wide range of foods, such as rice, beans, wine, beer, coffee, cocoa, dried fruit, and animal products. OTA is categorized by the International Agency for Research on Cancer (IARC) as a group 2B possible human carcinogen. It is hepatotoxic, teratogenic, immunosuppressive, nephrotoxic, and nephrocarcinogenic [1,2]. A number of countries have moved to establish regulatory limits on OTA in food products destined for human consumptions [3]. For instance, the European Commission has imposed regulatory limits on OTA in corn and corn products. A maximum of 5 μg/kg for natural corn grain, 3 μg/kg for all other corn products destined for direct human consumption, and 0.5 μg/kg for baby food and corn-based products intended for young children is allowed [4].

To safeguard human health against the food safety risks associated with OTA, advanced, sensitive, and accurate analytical methods are required for its detection and quantification [4]. Instrument-based methods like HPLC connected to a fluorescence detector (HPLC/FLD) and liquid chromatography/mass spectrometry (LC/MS) are some of the most widely-used mycotoxin detection techniques. While instrument-based methods offer precision and reliability, compared to newer analytic techniques, they have some weaknesses: they are costly, require a certain level of expertise to operate them, and are not suitable for on-site use [5,6]. To overcome these drawbacks, immunoassays have recently gained popularity as an alternative to the above-described methods. Based on the binding of antigen to antibody, immunoassay-based techniques are cheap, simple, and sensitive [7,8].

Certain immunoassay techniques, such as ELISA, require tedious and time-consuming assay development [8]. Alternatively, FPIA is a simple and user-friendly immunoassay as it does not require tedious and time-consuming pre-analytical steps [9,10]. FPIA is a widely used homogeneous-based immunoassay with simple and rapid operational procedures. Currently, the method is widely applied in the monitoring of small molecules in variety of matrices [11–13]. Fluorescence polarization (FP) is commonly used to excite fluorescent molecules with polarized light in a vertical direction, and then measure the fluorescence intensity Iv and Ip of polarized light emitted in the vertical and horizontal directions, respectively. FP = (IV-IP)/(Iv + Ip), where FP is a dimensionless quantity, and the unit is usually expressed in mP. The principle of fluorescence polarization detection is based on the different sizes of the fluorescence molecules and the different intensities of the fluorescence polarization signal [14].

The use of FPIA to detect mycotoxins such as aflatoxins (AFTs), fumonisins (FBs), deoxynivalenol (DON), ochratoxin A (OTA), zearalenone (ZEN), and HT-2 and T-2 toxins in various matrices as reviewed by Maragos [4]. Additionally, Li et al. reported the development of a multiplexed FPIA for the simultaneous determination of deoxynivalenol, T-2 toxin, and fumonisin in maize samples. With regards to OTA, an OTA-ethylenediamine fluorescence (EDF) conjugate-based FPIA with a limit of detection (LOD) of 0.3 ng/mL of OTA in unpolished rice was reported [15]. In this study, we synthesized four tracers, among which the OTA-AMF tracer was chosen for further optimization to improve the detection sensitivity.

In this work, four new tracers with different fluorophores were synthesized. Based on fluorescence intensity, we selected two apparently better tracers for the subsequent experiments. Two FPIAs were then researched for the determination of OTA in buffer by optimizing the reaction conditions. Based on the optimum tracers and sensitive antibody against OTA, we successfully built and applied a simple, fast, and sensitive FPIA for the detection of OTA in rice. Based on optimal conditions, we further validated the results that were obtained by FPIA using HPLC as a reference method. The developed FPIA, as a result, is a promising method for the rapid analysis of OTA-contaminated rice samples.

2. Results and Discussions

2.1. The Principle of FPIA

The principle of this type of immunoassays is based on the competition between native mycotoxins in the sample and the mycotoxin-labeled tracer for the monoclonal antibody (mAb) [16]. The addition of the tracer to the mAb influences the tracer molecule activation and enhances the FP value [17]. The quantity of synthetic tracer is inversely proportional to the amount of free mycotoxin that exists in the sample; consequently, the analyte concentration inversely correlates with the polarization value. Specifically, the small molecule to be measured is labeled with a fluorescent substance capable of generating polarized light and the change of FP value before and after the fluorescence marker is combined with specific antibody is measured (δFP = FPbind-FPfree). Then, a standard curve is established to achieve quantitative detection of the small molecule to be measured.

2.2. Preparation of Monoclonal Antibody

The 1H2 cells reached logarithmic phase four days after resuscitation. Ascites were collected 10 days after injection. The purification effect of the monoclonal antibody was determined with SDS-PAGE (12% separation gel, 5% spacer gel). The purified immunoglobulin G antibody has only two main bands, heavy chain and light chain, indicating that the purification method can remove the heteroprotein in ascites. The relationship between the molecular weight of the protein marker and the mobility of the heavy and light chains was obtained. The molecular weights of heavy and light chains were about 50 and 25 kDa, respectively (Figure 1). The purified antibodies proved to be capable of meeting the requirements for the next experiment.

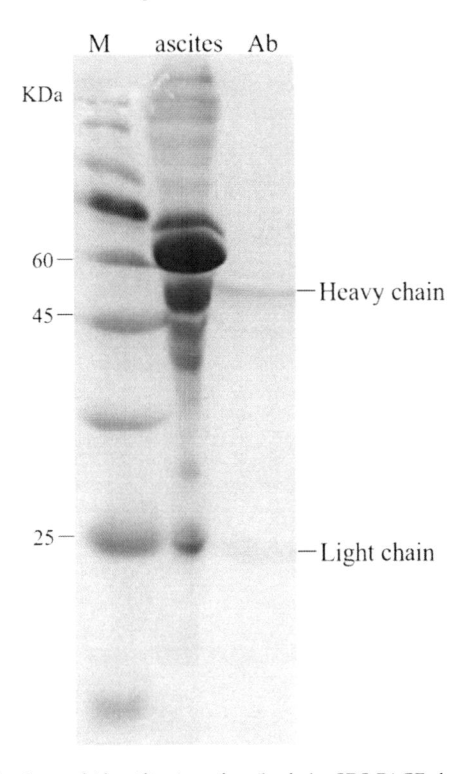

Figure 1. Purification and identification of antibody by SDS-PAGE electrophoretogram.

2.3. Synthesis of the OTA-FL Tracer

The carboxyl groups in OTA are inactive; for their activation, we used N, N′-dicyclohexylcarbodiimide and N-hydroxysuccinimide in an aprotic solvent medium. In this research, four typical kinds of dyes with different Ex/Em wavelengths (HDF, BDF, AMF, and EDF) were selected for covalent conjugation to OTA.

After preliminary purification through thin layer chromatography (TLC), principal bands of OTA-AMF, OTA-BDF, OTA-EDF, and OTA-HDF were collected. The molecular weights of OTA-AMF, OTA-BDF, OTA-EDF, and OTA-HDF were 747.15, 863.33, 837.29, and 891.38, respectively (Figure 2). The mass spectra [M+] ion peaks were 747.45, 863.45, 835.45, and 891.50, respectively, which are consistent with the molecular weights of the target compounds (Figure 3). Dye-labelled tracers were primarily designed to bind to the specific monoclonal antibodies to determine whether the OTA-AMF, OTA-BDF, OTA-EDF, and OTA-HDF could provide satisfactory results. All the tracers induced a significant rise in FP signals before and after the addition of saturated quantities of mAbs (Figure 4). The δFP values of the tracers ranged from 5 to 207 mP, which are adequate for application in FPIA reaction progress monitoring. This proved that the dyes were successfully conjugated to the corresponding mycotoxin [18]. To improve the detection sensitivity, the OTA-AMF that had the highest δFP value was chosen for further optimization.

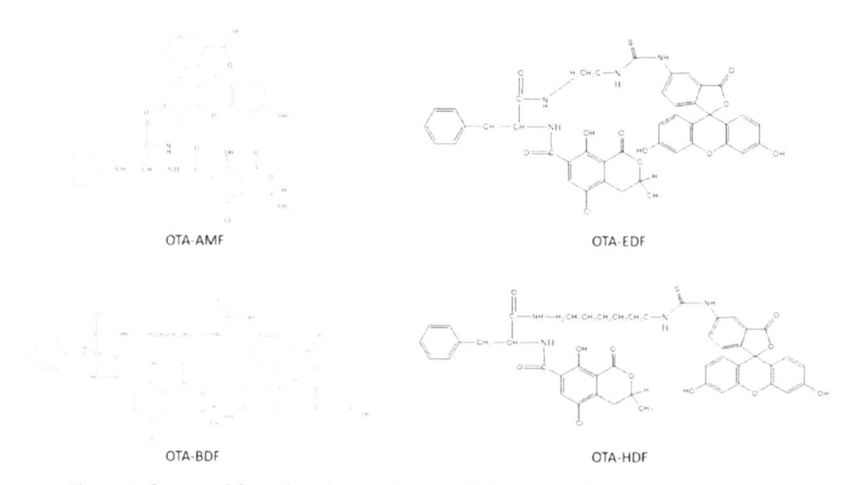

Figure 2. Structural formulas of tracer OTA-AMF, OTA-BDF, OTA-EDF, and OTA-HDF.

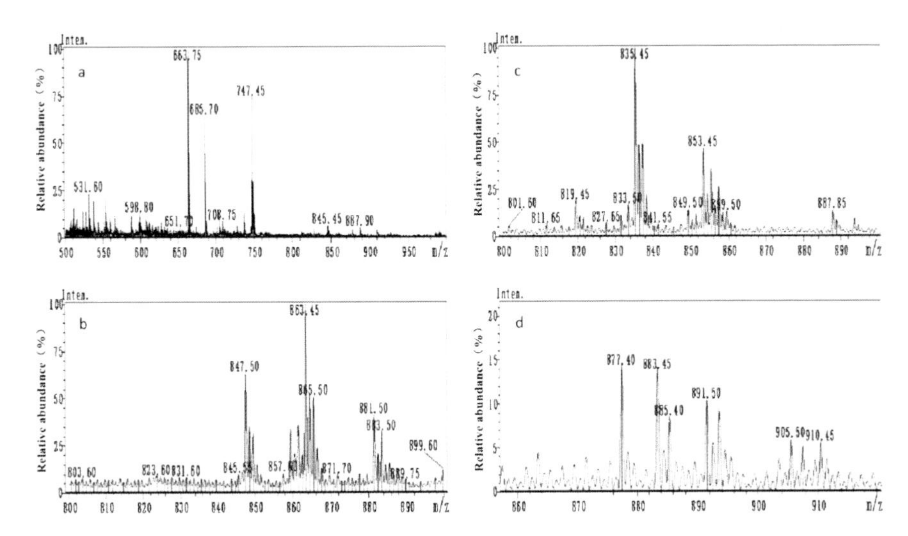

Figure 3. Mass spectra of (a) OTA-AMF, (b) OTA-BDF, (c) OTA-EDF, and (d) OTA-HDF.

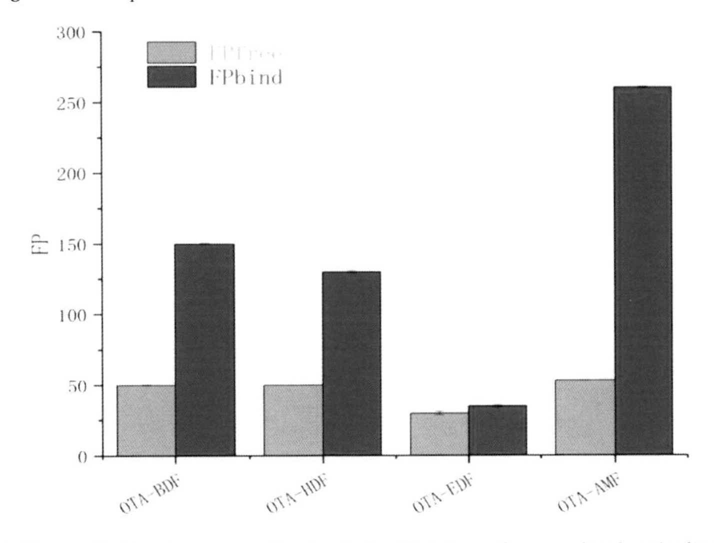

Figure 4. The result of four tracers combined with the diluted specific monoclonal antibodies (n = 3).

2.4. Optimization of the FPIA

The influence of methanol concentration and reaction time on the assay's performance was studied to evaluate the applicability of the method. OTA is typically extracted from cereals with methanol. The dyes that were applied in this experiment were methanol-sensitive substances; various physicochemical statuses can importantly affect the cross-linkage between the antigens and antibodies [19,20]. As the methanol concentration rose from 0% to 40%, the half maximal inhibitory concentration (IC_{50}) of the experiment slightly decreased (Figure 5a). The reaction time was studied at a concentration of 40% methanol. The best level of sensitivity was attained at 0–10 min; the IC_{50} values increased with the increase of the assay time (Figure 5b). Therefore, 40% methanol and 20 min reaction time were regarded as the optimum assay conditions and were applied in the subsequent experiments.

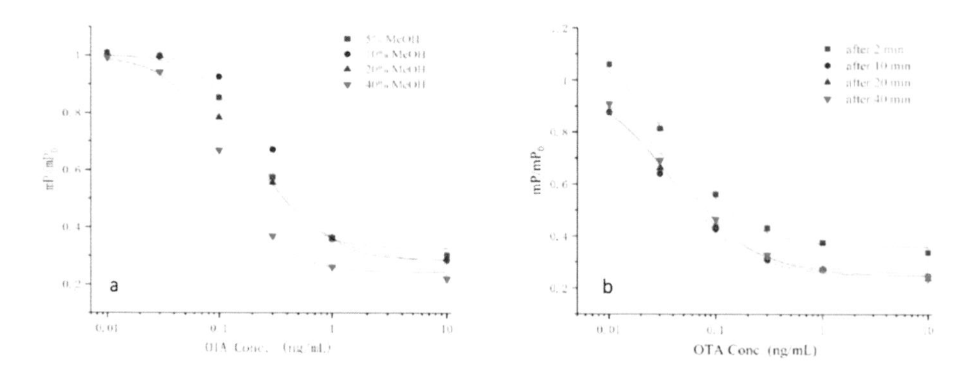

Figure 5. Normalized correction curves of the optimum FPIA implemented with OTA reference solutions with different methanol content (**a**) and reaction time (**b**) (n = 3).

Based on the optimum conditions, we established the OTA calibration curve within the concentration range of 0.03–0.78 ng/mL. When the correlation coefficient was equivalent to 0.996, the limit of detection (LOD) and IC_{50} were 0.02 and 0.09 ng/mL, respectively (Figure 6). Due to the dilution ratio of rice being 20:1, the linear detection range of this method for rice in the actual detection was 0.60–15.60 ng/mL.

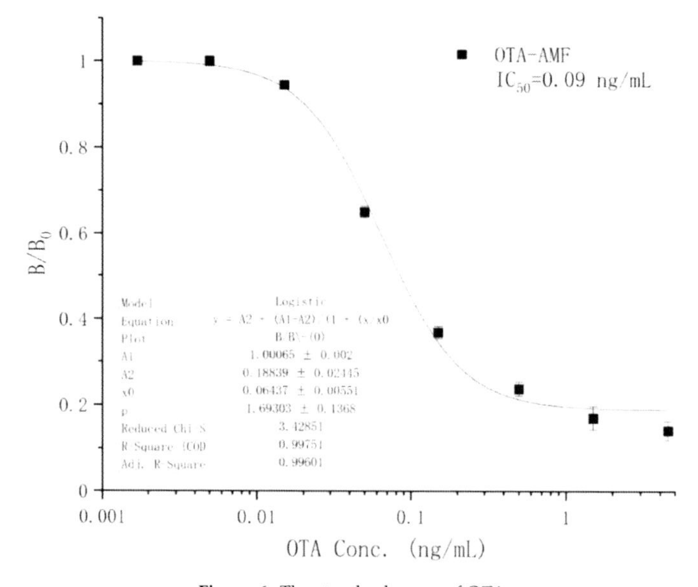

Figure 6. The standard curve of OTA.

2.5. Evaluation of FPIA

The recovery, intra-assay, and inter-assay tests were carried out to assess the performance of the FPIA method. The recovery tests in rice samples spiked with OTA standards of 0.5, 5, and 50 μg/kg concentration (Table 1) yielded an average recovery range of 70–110% with a relative standard deviation of equal to or less than 20%. The recovery and reproducibility values of this method proved that it fits the mycotoxin screening and testing methods' validation and performance criteria standard set by the EU for the control of mycotoxin in food products [4].

Table 1. Relative standard deviations and average recoveries of OTA from spiked rice samples obtained by FPIA.

Assay	Spiking Level (μg/kg)	Recovery (%)	CV (%)
Intra-assay	0.5	110	5.6
	5	72.5	6.2
	50	80.1	8.5
Inter-assay	0.5	113.2	6.2
	5	76.8	8.4
	50	80.4	9.8

2.6. FPIA Screening and HPLC Analysis of Blind Samples

A set of 10 naturally OTA-contaminated rice samples with a contamination levels ranging from 0.98 to 14.6 ng/mL (through HPLC analysis) was analyzed by both an immunoaffinity column clean-up with HPLC and the developed FPIA assay for comparison (Table 2). Linear regression of recovery data showed a good correlation ($r^2 = 0.9966$, Figure 7). An analysis of variance (ANOVA) estimation proved that OTA concentrations results obtained through the FPIA method were a good forecast of the expected values as tested by HPLC ($p < 0.0001$). Data from the two methods showed a good level of correlation and concurred with the spiked quantities, proving the practical applicability of the developed method.

Table 2. Quantitative determination of OTA in practical samples with FPIA and HPLC methods.

Sample No.	Ochratoxin A Determined (ng g^{-1}) FPIA	HPLC
1	15.1 ± 0.50	14.6 ± 0.31
2	1.5 ± 0.02	0.98 ± 0.05
3	2.1 ± 0.03	1.9 ± 0.04
4	10.5 ± 0.31	8.5 ± 0.28
5	2.6 ± 0.15	2.2 ± 0.10
6	5.0 ± 0.13	4.5 ± 0.12
7	3.5 ± 0.06	3.06 ± 0.03
8	15.0 ± 0.51	12.6 ± 0.27
9	10.8 ± 0.62	9.1 ± 0.16
10	2.6 ± 0.04	2.1 ± 0.06

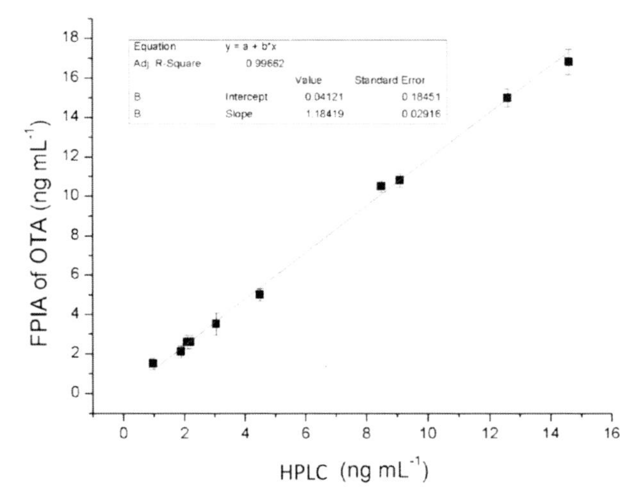

Figure 7. Correlation analysis between HPLC method and the developed FPIA assy.

3. Conclusions

A sensitive and accurate FPIA analytical method based on a new tracer (OTA-AMF) was developed and optimized for the rapid detection and quantification of OTA in rice and maize samples. The method detected OTA in concentrations ranging from 0.03–0.78 ng/mL and reached an IC_{50} value of 0.09 ng/mL. Compared to other immunoassays (Table 3), the FPIA method has better sensitivity. Apart from the purification procedure, the whole analytical process was completed within 10 min showing its practical applicability as a rapid screening method. The precision of the method was proven by comparing the results of OTA levels in rice and maize samples obtained by HPLC and results obtained by the FPIA. Furthermore, the proposed technique is inexpensive, user-friendly, and suitable for on-site use. The method is suitable for screening and quantitative determination of OTA in rice and maize at levels lower the EU regulatory limits for OTA in cereal grains and can serve as an alternative to instrument-based methods.

Table 3. Comparing the sensitivity with other immuno-assay methods.

Author	Year	Antibody	Experiment Method	Sample	Sensitivity (IC_{50}, ng/mL)	LOD (ng/mL)
This paper	2020	mAb	FPIA	rice	0.09	0.02
Becheva [21]	2020	F(ab')2	FIA	milk	a	0.08
Beloglazova [22]	2020	mAb	Flow-through Immunoassay	feed	10	
Wang [23]	2020	Nb	FIA	food	0.46	0.12
Chen [7]	2019	mAb	FPIA	yoghurt	9.32	0.82
Zhang [24]	2019	Nb	ELISA	cereals	97	–
Rehmat [25]	2019	mAb	SPR immunoassay	coffee		3.8
Qin [26]	2019	mAb	ELISA	nutmeg	0.146	0.031
Machado [27]	2018	mAb	capillary micro-fluidic immunoassay	feed		40
Tang [28]	2018	Nb	one-step immunoassay	cereal	5	
Soares [29]	2018	mAb	FIA			1
Sun [30]	2018	Nb	ELISA	rice	0.57	0.059
Liu [31]	2017	Nb	ELISA	cereal	0.64	
Lippolis [32]	2017	mAb	FPIA	rye		0.6
Majdinasab [33]	2015	mAb	TRFICA	agro-product		1
Lippolis [4]	2014	mAb	FPIA	wheat	0.48	0.8
Li [34]	2013	mAb	immunochromatographic assay	agro-food		0.5
Bondarenko [35]	2012	mAb	FPIA	grain		10
Zezza [36]	2009	mAb	FPIA	red wine		0.7

[a] The data were not detected or shown in the paper. mAb: monoclonal antibody. Nb: nanobody. FIA: fluoroimmunoassay. ELISA: enzyme-linked immunosorbent assay. TRFICA: time-resolved fluorescent immunochromatographic assay.

4. Materials and Methods

4.1. Reagents and Chemicals

Ochratoxin A (OTA), N,N-dimethylformamide (DMF), 1-ethyl-3-(3-dimethylaminopropyl) carbodiimide (EDC), N-hydroxysuccinimide (NHS), octanoic acid, ammonium sulfate, dimethyl sulfoxide (DMSO), and Freund's incomplete adjuvant liquid (FICA) were bought from Sigma-Aldrich. Modified RPMI medium, HEPES, and penicillin-streptomycin solution were obtained from GE Healthcare-Hyclone. Fetal bovine serum was obtained from Gibco. Cell culture flask (75 cm^2) was obtained from CORNING. Dialysis tubing (MWCO 15 KDa) was obtained from Thermo-Scientific. The SDS-PAGE gel preparation kit was obtained from Biosharp Life Sciences. Thin-layer chromatography (TLC) plates were obtained from Darmstadt; the model specifications were silica gel 60, 1 mm, 20 × 20 cm, with fluorescent indicator. Fluorescein thiocarbamoyl ethylenediamine (EDF), fluorescein thiocarbamoyl butane diamine (BDF), fluorescein thiocarbamoyl hexame (HDF), and amino-methyl fluorescein (AMF) were provided by Sergei A. Eremin, a professor of Department of

Chemistry, Lomonosov Moscow State University. All FPIA experiments used sodium borate solution (BB, 0.05 M, pH = 9.0). Phosphate buffer solution (PBS, 0.01 M, pH = 7.4) was used for the mAb dialysis. All of the organic solvents and chemicals were reagent grade or above.

The fluorescence polarizer was a sentinel 200 FP portable unit (Diachemix, Grayslake, 1 L). Measurements of intensity and fluorescence polarization were executed applying the TDx/FLx Analyzer (Abbott, Irving, TX, USA) in an aided Photo Check method. TDx/FLx glass cuvettes were put into the particular turntable up to 10 at a time, then polarization (mP units) and fluorescence intensity (customary units) were measured. The total measurement time of 10 samples was about 7 min.

4.2. Animals and Cells

Female BALB/c mice(age 8–10 weeks), which were used for the production of antibodies, were purchased from the Institute of Biological Products of Hubei Province (Wuhan, China). OTA hybridoma cell strain 1H2 was developed in our laboratory [37].

4.3. Hybridoma Cell Culture and Antibody Preparation

OTA hybridoma cell strain 1H2 was taken out of a liquid nitrogen container, melted in a 37 °C water bath for 1 min, and washed with modified RPMI medium solution. The cells were resuspended with 1640 complete medium (RPMI medium modified: HEPES: penicillin-streptomycin, $V:V:V:V$ = 80:20:1:1), and transferred to a cell culture flask, then incubated in a constant temperature incubator (at 37 °C, CO_2 5%). Hybridoma cells were collected during the logarithmic phase, and injected into the BALB/c mice that had been treated with FICA. The number of cell injections was maintained at 2×10^6. Ascites were harvested with drainage needle about one week later.

The ascites were used to prepare pure OTA antibodies, which were purified by caprylic acid-ammonium sulfate precipitation [38]. Subsequently, the antibodies were freeze-dried in a vacuum, and stored at −20 °C for later use. The purity of monoclonal antibody was determined by the SDS-PAGE gel electrophoresis method.

4.4. Preparation of Four Different OTA-Fluorescein Tracers

OTA has an active group that can be coupled with commonly used fluorescein and directly reacted with FITC to become a tracer OTA-EDF (or OTA-BDF, OTA-AMF, and OTA-HDF) as previously reported [39,40]. In brief, 250 µL (80 µmol/mL) NHS and 250 µL (80 µmol/mL) EDC were dissolved in 0.2 mL DMF, then 2 mg (5 µmol) OTA was added. The solution was stirred for 2 h and subsequently reposed overnight at room temperature. For the synthesis of OTA-EDF, 2 mg (4 µmol) EDF was added into 200 µL (2 µmol) activated OTA solution, followed by 10 h incubation at room temperature. For OTA-BDF and OTA-HDF, 1 mg (2 µmol) BDF (or HDF) was added into 100 µL (1 µmol) activated OTA solution, followed by 10 h incubation at room temperature and an aliquot. For OTA-AMF, 0.5 mg (1 µmol) AMF was added into 100 µL (1 µmol) activated OTA solution after 10 h reaction at room temperature.

An aliquot of the mixture was separated and purified by TLC. The silica gel plates were activated at 110 °C for 30 min before use. Toluene ± acetic acid (99:1 *v*/*v*) was chosen as the spotting solvent. Toluene ± ethyl acetate ± 88% formic acid (6:3:1 *v*/*v*/*v*) was regarded as the suitable solvent for eluting samples on TLC plates. The TLC plates were examined visually under UV light at 365 nm [36].

The OTA-AMF and other tracers were dissolved in 2 mL and 1 mL methanol, respectively, for separation and purification. A primary band at $Rf\frac{1}{4}$ 0.9 and $Rf\frac{1}{4}$ 0.7 were collected and eluted with methanol. The methanol tracer solutions were stored at −20 °C.

Dye-labeled tracers were identified by mass spectrometry (SHIMADZU, LCMS-8060, liquid chromatograph-mass spectrometer). The instrument was corrected with raffinose before the experiment. The OTA-AMF, OTA-BDF, OTA-EDF, and OTA-HDF were diluted to 1 µg/mL with 1 mL methanol, and then injected into the mass spectrometer through autosampler. The parameters of the mass spectrum were adjusted to real-time at the beginning of data collection. The parameters of the compound ion mass spectrum were obtained and analyzed for their chemical structures.

4.5. Method of Fluorescence Polarization Immunoassay

The concentrations of OTA standard working solution were 0, 0.0034, 0.01, 0.03, 0.1, 0.3, 1, 3, and 9 ng/mL prepared with 10% methanol in deionized water. The FPIA was prepared by 0.1 M borate buffer (pH = 7.4); the antibody working fluid was based on attenuating OTA specific antibody (mAb) 1:36,000 in BB buffer. Glass culture tubes with specifications 10 × 75 mm (VWR Scientific) were used as test cuvettes. We added 500 μL antibody working solution into the tube, then 500 μL OTA-EDF (or OTA-BDF/OTA-AMF/OTA-HDF) working solution and mixed. The FP value was measured after 10 min of oscillate incubation at ambient temperature; the relative FP mean values (mP/mP_0) were used in the inhibition curve to standardize the FP value, where mP is the current FP value of different OTA concentrations and mP_0 is the value of blank-control (50 μL methanol-BB solution was used as the blank-control) [41]. The values of mP/mP_0 were plotted against OTA concentration [37]. For experiments to elucidate the reaction's kinetics, measurements were recorded for time intervals ranging from 3 s to 10 min at room temperature, unless otherwise noted. OTA content of naturally contaminated rice samples was approximated based on the OTA-PBS solution specification curve [42].

4.6. Sample Preparation

A total of 5 g of rice was extracted and vortex mixed with 25 mL 50% methanol for 30 min. Then, the supernatant fluid was first filtered with a double filter paper before it was further purified with a 0.22 μm filter membrane. Then, 250 μL of the filtrate was watered down with 750 μL deionized water, samples of which were analyzed with the FPIA method [43].

4.7. Comparison with HPLC Analysis

Samples were extracted with ochratoxin A immunoaffinity columns similar to those previously described. Briefly, rice samples were finely ground and homogenized and then mixed with 80% acetonitrile and half volume hexane by gently mixing. The extracts were filtered with a filter paper, and then the filtrate was collected and centrifuged. The bottom layer was evaporated to dry below a flow of nitrogen. After dilution with acetonitrile and PBS, the extracted samples were loaded into the IAC columns. OTA was eluted with 2% methanol/acetic acid solution, and then dry-evaporated. A 50 μL reconstituted sample (1 mL acetonitrile) was injected into the chromatograph [44,45]. The HPLC analysis was run applying a C18 column on a Waters Alliance 2695 chromatographic system in isocratic conditions at ambient temperature with the moving phase of CH_3CN:NH_3/NH_4Cl (20 mM, pH = 9.8) (*v/v* = 15:85); the column was acquired using Waters XTerra® (3 μm, 2.1 × 250 mm), the injection volume was 20 μL, and the flow velocity was 0.2 mL per minute; the FLD determination was acquired using a Waters 474 Scanning Fluorescence Detector (λex 380 nm, λem 440 nm; attenuation 32; gain 7 × 100; bandwidth 40 nm); the analyte holding time was 20 times the retention time, corresponding to the column void volume; no chemical compound could be used as an internal reference for the OTA extraction [46].

Author Contributions: Conceptualization, X.T., X.H. and Q.Z.; methodology, X.T. and X.H.; software, X.H.; validation, X.T., X.H. and Q.Z.; formal analysis, X.T. and X.H.; investigation, X.T. and X.H.; resources, Q.Z., P.L., X.Q., W.Z., J.J. and H.L.; data curation, X.T. and X.H.; writing—original draft preparation, X.H.; writing—review and editing, X.H., Q.Z. and A.J.; visualization, X.H.; supervision, X.T. and Q.Z.; project administration, Q.Z., P.L., W.Z., J.J. and H.L.; funding acquisition, Q.Z. and P.L. All authors have read and agreed to the published version of the manuscript.

Funding: This research was funded by the National Key R&D Program of China (2018YFC1602505), Agricultural Science and Technology Innovation Program of CAAS (Chinese Academy of Agricultural Sciences) (CAAS-ZDRW202011), the Natural Science Foundation of China (31801665)

Acknowledgments: The Fluorescein thiocarbamyl ethylenediamine (EDF), Fluorescein thiocarbamyl butane diamine (BDF), fluorescein thiocarbamyl hexame (HDF) and amino-methyl fluorescein (AMF) were provided by Sergei A. Eremin, a professor of Department of Chemistry, Lomonosov Moscow State University.

Conflicts of Interest: The authors declare no conflict of interest. We declare that we do not have any commercial or associative interest that represents a conflict of interest in connection with the work submitted.

References

1. Naohiko, A.; Promsuk, J.; Hitoshi, E. Molecular Mechanism of Ochratoxin A Transport in the Kidney. *Toxins* **2010**, *2*, 1381–1398.
2. Fazekas, B.; Tar, A.; Kovacs, M. Aflatoxin and ochratoxin A content of spices in Hungary. *Food Addit. Contam.* **2005**, *22*, 856–863. [CrossRef] [PubMed]
3. Anukul, N.; Vangnai, K.; Mahakarnchanakul, W. Significance of regulation limits in mycotoxin contamination in Asia and risk management programs at the national level. *J. Food Drug Anal.* **2013**, *21*, 227–241. [CrossRef]
4. Lippolis, V.; Pascale, M.; Valenzano, S.; Porricelli, A.C.R.; Suman, M.; Visconti, A. Fluorescence Polarization Immunoassay for Rapid, Accurate and Sensitive Determination of Ochratoxin A in Wheat. *Food Anal. Methods* **2013**, *7*, 298–307. [CrossRef]
5. Cigić, I.K.; Prosen, H. An overview of conventional and emerging analytical Methods for the determination of mycotoxins. *Int. J. Mol. Sci.* **2009**, *10*, 62–115. [CrossRef] [PubMed]
6. Vettorazzi, A.; González-Peñas, E.; De Cerain, A.L. Ochratoxin A kinetics: A review of analytical methods and studies in rat model. *Food Chem. Toxicol.* **2014**, *72*, 273–288. [CrossRef] [PubMed]
7. Chen, Y.; He, Q.; Shen, D.; Jiang, Z.; Eremin, S.A.; Zhao, S. Fluorescence polarization immunoassay based on a new monoclonal antibody for the detection of the Diisobutyl phthalate in Yoghurt. *Food Control* **2018**, *105*, 38–44. [CrossRef]
8. Zhang, X.; Song, M.; Yu, X.; Wang, Z.; Ke, Y.; Jiang, H.; Li, J.; Shen, J.; Wen, K. Development of a new broad-specific monoclonal antibody with uniform affinity for aflatoxins and magnetic beads-based enzymatic immunoassay. *Food Control* **2017**, *79*, 309–316. [CrossRef]
9. Huang, P.; Zhao, S.; Eremin, S.A.; Zheng, S.; Lai, D.; Chen, Y.; Guo, B. A fluorescence polarization immunoassay method for detection of the bisphenol A residue in environmental water samples based on a monoclonal antibody and 4′-(aminomethyl) fluorescein. *Anal. Methods* **2015**, *7*, 4246–4251. [CrossRef]
10. Zhang, X.; Tang, Q.; Mi, T.; Zhao, S.; Wen, K.; Guo, L.; Mi, J.; Zhang, S.; Shi, W.; Shen, J.; et al. Dual-wavelength fluorescence polarization immunoassay to increase information content per screen: Applications for simultaneous detection of total aflatoxins and family zearalenones in maize. *Food Control* **2018**, *87*, 100–108. [CrossRef]
11. Lea, W.A.; Simeonov, A. Fluorescence polarization assays in small molecule screening. *Expert Opin. Drug Discov.* **2010**, *6*, 17–32. [CrossRef]
12. Maragos, C. Fluorescence polarization immunoassay of mycotoxins: A review. *Toxins* **2009**, *1*, 196–207. [CrossRef]
13. Rossi, A.M.; Taylor, C.W. Analysis of protein-ligand interactions by fluorescence polarization. *Nat. Protoc.* **2011**, *6*, 365–387. [CrossRef]
14. Lakowicz, J.R. *Principles of Fluorescence Spectroscopy*; Springer: Berlin/Heidelberg, Germany, 2006; Volume 13, p. 029901.
15. Park, J.H.; Chung, D.H.; Lee, I.S. Application of Fluorescence Polarization Immunoassay for the Screening of Ochratoxin A in Unpolished Rice. *J. Life Sci.* **2006**, *16*, 1006–1013.
16. Maragos, C.; Jolley, M.E.; Plattner, R.D.; Nasir, M.S. Fluorescence polarization as a means for determination of fumonisins in maize. *J. Agric. Food Chem.* **2001**, *49*, 596–602. [CrossRef]
17. Shim, W.B.; Kolosova, A.; Kim, Y.J.; Yang, Z.Y.; Park, S.J.; Eremin, S.A.; Lee, I.-S.; Chung, D.H. Fluorescence polarization immunoassay based on a monoclonal antibody for the detection of ochratoxin A. *Int. J. Food Sci. Technol.* **2004**, *39*, 829–837. [CrossRef]
18. Li, C.; Wen, K.; Mi, T.; Zhang, X.; Zhang, H.; Zhang, S.; Shen, J.; Wang, Z. A universal multi-wavelength fluorescence polarization immunoassay for multiplexed detection of mycotoxins in maize. *Biosens. Bioelectron.* **2016**, *79*, 258–265. [CrossRef]

19. Mortensen, G.K.; Strobel, B.W.; Hansen, H.C.B. Determination of zearalenone and ochratoxin A in soil. *Anal. Bioanal. Chem.* **2003**, *376*, 98–101. [CrossRef]
20. Peters, J.; Thomas, D.S.; Boers, E.; De Rijk, T.; Berthiller, F.; Haasnoot, W.; Nielen, M.W.F. Colour-encoded paramagnetic microbead-based direct inhibition triplex flow cytometric immunoassay for ochratoxin A, fumonisins and zearalenone in cereals and cereal-based feed. *Anal. Bioanal. Chem.* **2013**, *405*, 7783–7794. [CrossRef]
21. Becheva, Z.R.; Atanasova, M.K.; Ivanov, Y.L.; Godjevargova, T.I. Magnetic Nanoparticle-Based Fluorescence Immunoassay for Determination of Ochratoxin A in Milk. *Food Anal. Methods* **2020**, *2020*, 1–11.
22. Beloglazova, N.V.; Graniczkowska, K.; Njumbe Ediage, E.; Averkieva, O.; De Saeger, S. Sensitive Flow-through Immunoassay for Rapid Multiplex Determination of Cereal-borne Mycotoxins in Feed and Feed Ingredients. *J. Agric. Food Chem.* **2017**, *65*, 7131–7137. [CrossRef] [PubMed]
23. Wang, X.; Wang, Y.; Wang, Y.; Chen, Q.; Liu, X. Nanobody-alkaline phosphatase fusion-mediated phosphate-triggered fluorescence immunoassay for ochratoxin a detection. *Spectrochim. Acta Part A Mol. Biomol. Spectrosc.* **2019**, *226*, 117617. [CrossRef] [PubMed]
24. Zhang, C.; Zhang, Q.; Tang, X.; Zhang, W.; Li, P.-W. Development of an Anti-Idiotypic VHH Antibody and Toxin-Free Enzyme Immunoassay for Ochratoxin A in Cereals. *Toxins* **2019**, *11*, 280. [CrossRef] [PubMed]
25. Rehmat, Z.; Mohammed, W.S.; Sadiq, M.B.; Somarapalli, M.; Anal, A.K. Ochratoxin A detection in coffee by competitive inhibition assay using chitosan-based surface plasmon resonance compact system. *Colloids Surf. B Biointerfaces* **2019**, *174*, 569–574. [CrossRef]
26. Qin, L.; Zhang, L.; Jiang, J.-Y.; Wang, C.-J.; Dou, X.-W.; Wan, L.; Yang, M.-H. High-throughput screening of ochratoxin A in Chinese herbal medicines using enzyme-linked immunoassay. *China J. Chin. Mater. Med.* **2019**, *44*, 5072–5077.
27. Machado, J.M.D.; Soares, R.R.G.; Chu, V.; Conde, J.P. Multiplexed capillary microfluidic immunoassay with smartphone data acquisition for parallel mycotoxin detection. *Biosens. Bioelectron.* **2018**, *99*, 40–46. [CrossRef]
28. Tang, Z.; Wang, X.; Lv, J.; Hu, X.; Liu, X. One-step detection of ochratoxin A in cereal by dot immunoassay using a nanobody-alkaline phosphatase fusion protein. *Food Control* **2018**, *92*, 430–436. [CrossRef]
29. Soares, R.; Santos, D.R.; Pinto, I.F.; Azevedo, A.M.; Aires-Barros, M.R.; Chu, V.; Conde, J.P. Multiplexed microfluidic fluorescence immunoassay with photodiode array signal acquisition for sub-minute and point-of-need detection of mycotoxins. *Lab Chip* **2018**, *18*, 1569–1580. [CrossRef] [PubMed]
30. Sun, Z.; Wang, X.; Chen, Q.; Yun, Y.-H.; Tang, Z.; Liu, X. Nanobody-Alkaline Phosphatase Fusion Protein-Based Enzyme-Linked Immunosorbent Assay for One-Step Detection of Ochratoxin A in Rice. *Sensors* **2018**, *18*, 4044. [CrossRef] [PubMed]
31. Liu, X.; Tang, Z.; Duan, Z.; He, Z.; Shu, M.; Wang, X.; Gee, S.J.; Hammock, B.D.; Xu, Y. Nanobody-based enzyme immunoassay for ochratoxin A in cereal with high resistance to matrix interference. *Talanta* **2017**, *164*, 154–158. [CrossRef]
32. Vincenzo, L.; Anna, P.; Marina, C.; Michele, S.; Sandro, Z.; Michelangelo, P. Determination of Ochratoxin A in Rye and Rye-Based Products by Fluorescence Polarization Immunoassay. *Toxins* **2017**, *9*, 305.
33. Majdinasab, M.; Sheikh-Zeinoddin, M.; Soleimanian-Zad, S.; Li, P.; Zhang, Q.; Li, X.; Tang, X.; Li, J. A reliable and sensitive time-resolved fluorescent immunochromatographic assay (TRFICA) for ochratoxin A in agro-products. *Food Control* **2015**, *47*, 126–134. [CrossRef]
34. Li, X.; Li, P.; Zhang, Q.; Li, R.; Zhang, W.; Zhang, Z.; Ding, X.; Tang, X. Multi-component immunochromatographic assay for simultaneous detection of aflatoxin B1, ochratoxin A and zearalenone in agro-food. *Biosens. Bioelectron.* **2013**, *49*, 426–432. [CrossRef]
35. Bondarenko, A.P.; Eremin, S.A. Determination of zearalenone and ochratoxin a mycotoxins in grain by fluorescence polarization immunoassay. *J. Anal. Chem.* **2012**, *67*, 790–794. [CrossRef]
36. Zezza, F.; Longobardi, F.; Pascale, M.; Eremin, S.A.; Visconti, A. Fluorescence polarization immunoassay for rapid screening of ochratoxin A in red wine. *Anal. Bioanal. Chem.* **2009**, *395*, 1317–1323. [CrossRef] [PubMed]
37. Li, X.; Li, P.; Zhang, Q.; Zhang, Z.; Li, R.; Zhang, W.; Ding, X.; Chen, X.; Tang, X. A Sensitive Immunoaffinity Column-Linked Indirect Competitive ELISA for Ochratoxin A in Cereal and Oil Products Based on a New Monoclonal Antibody. *Food Anal. Methods* **2013**, *6*, 1433–1440. [CrossRef]
38. McKinney, M.M.; Parkinson, A. A simple, non-chromatographic procedure to purify immunoglobulins from serum and ascites fluid. *J. Immunol. Methods* **1987**, *96*, 271–278. [CrossRef]

39. Jameson, D.M.; Ross, J.A. Fluorescence Polarization/Anisotropy in Diagnostics and Imaging. *Chem. Rev.* **2010**, *110*, 2685–2708. [CrossRef]
40. Santos, E.A.; Vargas, E.A. Immunoaffinity column clean-up and thin layer chromatography for determination of ochratoxin A in green coffee. *Food Addit. Contam.* **2002**, *19*, 447–458. [CrossRef]
41. Chen, Y.; Cui, X.; Wu, P.; Jiang, Z.; Jiao, L.; Hu, Q.; Eremin, S.A.; Zhao, S. Development of a Homologous Fluorescence Polarization Immunoassay for Diisobutyl Phthalate in Romaine Lettuce. *Food Anal. Methods* **2016**, *10*, 449–458. [CrossRef]
42. Maragos, C.M.; Plattner, R.D. Rapid Fluorescence Polarization Immunoassay for the Mycotoxin Deoxynivalenol in Wheat. *J. Agric. Food Chem.* **2002**, *50*, 1827–1832. [CrossRef] [PubMed]
43. Tang, X.; Li, P.; Zhang, Z.; Zhang, Q.; Guo, J.; Zhang, W. An ultrasensitive gray-imaging-based quantitative immunochromatographic detection method for fumonisin B1 in agricultural products. *Food Control* **2017**, *80*, 333–340. [CrossRef]
44. Drabent, R.; Pliszka, B.; Olszewska, T. Fluorescence properties of plant anthocyanin pigments. I. Fluorescence of anthocyanins in Brassica oleracea L. extracts. *J. Photochem. Photobiol. B Biol.* **1999**, *50*, 53–58. [CrossRef]
45. Varelis, P.; Leong, S.-L.L.; Hocking, A.D.; Giannikopoulos, G.; Hedén, S.-L. Quantitative analysis of ochratoxin A in wine and beer using solid phase extraction and high performance liquid chromatography–fluorescence detection. *Food Addit. Contam.* **2006**, *23*, 1308–1315. [CrossRef]
46. Toscani, T.; Moseriti, A.; Dossena, A.; Dall'Asta, C.; Simoncini, N.; Virgili, R. Determination of ochratoxin A in dry-cured meat products by a HPLC–FLD quantitative method. *J. Chromatogr. B Anal. Technol. Biomed. Life* **2007**, *855*, 242–248. [CrossRef]

Publisher's Note: MDPI stays neutral with regard to jurisdictional claims in published maps and institutional affiliations.

Article

Determination of Zearalenone and Trichothecenes, Including Deoxynivalenol and Its Acetylated Derivatives, Nivalenol, T-2 and HT-2 Toxins, in Wheat and Wheat Products by LC-MS/MS: A Collaborative Study

Annalisa De Girolamo, Biancamaria Ciasca, Michelangelo Pascale and Veronica M.T. Lattanzio *

Institute of Sciences of Food Production, National Research Council of Italy, via Amendola 122/O, 70126 Bari, Italy; annalisa.degirolamo@ispa.cnr.it (A.D.G.); biancamaria.ciasca@ispa.cnr.it (B.C.); michelangelo.pascale@ispa.cnr.it (M.P.)

* Correspondence: veronica.lattanzio@ispa.cnr.it

Received: 12 November 2020; Accepted: 8 December 2020; Published: 10 December 2020

Abstract: An analytical method for the simultaneous determination of trichothecenes—namely, nivalenol (NIV), deoxynivalenol (DON) and its acetylated derivatives (3- and 15-acetyl-DON), T-2 and HT-2 toxins—and zearalenone (ZEN) in wheat, wheat flour, and wheat crackers was validated through a collaborative study involving 15 participants from 10 countries. The validation study, performed within the M/520 standardization mandate of the European Commission, was carried out according to the IUPAC (International Union of Pure and Applied Chemistry) International Harmonized Protocol. The method was based on mycotoxin extraction from the homogenized sample material with a mixture of acetonitrile-water followed by purification and concentration on a solid phase extraction column. High-performance liquid chromatography coupled with tandem mass spectrometry was used for mycotoxin detection, using isotopically labelled mycotoxins as internal standards. The tested contamination ranges were from 27.7 to 378 μg/kg for NIV, from 234 to 2420 μg/kg for DON, from 18.5 to 137 μg/kg for 3-acetyl-DON, from 11.4 to 142 μg/kg for 15-acetyl-DON, from 2.1 to 37.6 μg/kg for T-2 toxin, from 6.6 to 134 μg/kg for HT-2 toxin, and from 31.6 to 230 μg/kg for ZEN. Recoveries were in the range 71–97% with the lowest values for NIV, the most polar mycotoxin. The relative standard deviation for repeatability (RSD_r) was in the range of 2.2–34%, while the relative standard deviation for reproducibility (RSD_R) was between 6.4% and 45%. The HorRat values ranged from 0.4 to 2.0. The results of the collaborative study showed that the candidate method is fit for the purpose of enforcing the legislative limits of the major *Fusarium* toxins in wheat and wheat-based products.

Keywords: trichothecenes; zearalenone; *Fusarium* toxins; wheat; liquid chromatography–mass spectrometry; official control; collaborative study

Key Contribution: A method for the determination of trichothecenes and zearalenone by high-performance liquid chromatography coupled with tandem mass spectrometry after clean-up by solid phase extraction has been validated in a collaborative study involving 15 participant laboratories following the provisions given in the AOAC/IUPAC International Harmonized Protocol. Results show that the candidate method is fit for the purpose of enforcing the legislative limits of the major *Fusarium* toxins in wheat and wheat-based products.

1. Introduction

The mycotoxins nivalenol (NIV); deoxynivalenol (DON) and its acetyl derivatives, 3-acetyl DON (3-AcDON) and 15-acetyl DON (15-AcDON); T-2 toxin (T-2) and its metabolite HT-2 toxin (HT-2);

and zearalenone (ZEN) are produced by various *Fusarium* species. Cereals such as wheat, maize, barley, oats, rye, and relevant derived products are most likely to be affected [1]. The toxic effects of *Fusarium* toxins on human and animal species have been extensively studied [2–5]. Because of its ability to induce acute vomiting in pigs, DON has also been given the trivial name "vomitoxin". Its acute effects in humans are similar to those observed in animals. The most common effects of long-term dietary exposure to DON are weight gain suppression and anorexia [2]. Similarly, toxicity studies have shown that NIV has anorectic effects upon short-term exposure, as well as immunotoxic and hematotoxic effects [5]. T-2 and HT-2 are known to inhibit the synthesis of protein, DNA, and RNA, and to have immunosuppressive and cytotoxic effects [4]. Estrogenic activity is the critical mode of action of ZEN and its main reductive metabolites. Adverse effects have been reported on the reproductive tract either in male and female animals, including testosterone synthesis, spermatogenesis, fertility, and embryo survival [3]. According to the available toxicological data concerning carcinogenicity in humans, ZEN, DON, and NIV were included by the World Health Organization's (WHO) International Agency for Research on Cancer (IARC) in the Group 3 as not classifiable regarding their carcinogenicity to humans [6].

To ensure the safety of cereals and derived products, the European Commission (EC) set maximum levels for mycotoxins, including DON and ZEN [7]. Regarding T-2 and HT-2, indicative levels to stimulate data collection for the sum of these *Fusarium* mycotoxins in cereals and cereal-derived products were published in the Commission Recommendation 2013/165/EU [8]. When evaluating the risks to human and animal health related to the presence of DON, its acetylated derivatives (3- and 15-AcDON) and the modified form DON-3-glucoside, as well as T-2 and HT-2 in food and feed, the European Food Safety Authority (EFSA) recommended the interlaboratory validation and standardization of LC–MS/MS methodologies for the simultaneous quantification of DON and its derivatives and analytical methods with an appropriate sensitivity for T-2 and HT-2 toxins in food and feed commodities. In addition, the collection of analytical data on the co-occurrence of the above-mentioned toxins was requested [2,9]. Even though not regulated yet, available studies on the occurrence and toxicity of NIV have been evaluated by the EFSA, highlighting the need for validated methods suitable for the determination of NIV in the low μg/kg range to provide reliable data for risk assessment [5,10].

The establishment of standardized methods of analysis is of the utmost importance to guarantee a uniform application of the European legislation in all member states and contribute to maintaining a high level of food and feed safety. The results of collaborative studies and proficiency tests (PTs) made available in the last decade have provided evidence that liquid chromatography-mass spectrometry (LC-MS)-based methods of analysis for (multiple) mycotoxins in food/feed can be actually fit for the purpose of the enforcement of legislative limits [11–13]. At the EU level, standardized methods of analysis are established by the European Committee for Standardization (CEN). The first standard LC-MS-based methodologies for the determination of mycotoxins in foods were published in 2017 by the CEN and cover ZEN [14] and T-2/HT-2 toxins [15] determination. These two methods were validated within the M/520 standardization mandate [16] by which the Commission invited CEN to establish European Standards/Technical Specifications that provide standardized methods of analysis for mycotoxins in food. Six of the 11 methods of analysis listed in this mandate were specifically requested to be based on LC-MS.

The work described herein addresses item 4 of the standardization mandate, aiming at optimizing and validating an analytical method for the simultaneous determination of NIV, DON and its acetyl derivatives (3-AcDON and 15-AcDON), T-2, HT-2, and ZEN in cereals and cereal products by liquid chromatography-tandem mass spectrometry (LC-MS/MS). According to the EC sector information and statistics on cereals (https://ec.europa.eu/info/food-farming-fisheries/plants-and-plant-products/plant-products/cereals_en), more than 50% of the cereals grown in the EU are wheat. It was therefore agreed within the working group (CEN TC275/WG5 biotoxins) to select wheat and wheat products as representative target commodities for the validation study. The results of the full collaborative study, involving seventeen international laboratories, are reported together with information relevant to the preparation and characterization of the test materials, the protocol of analysis, and the statistical analysis of the results. The derived method has been recently adopted as a CEN standard [17].

2. Results and Discussion

2.1. Pre-Trial Results

Prior to the full validation study, laboratories had to participate in a pre-trial study. The aims of the pre-trial were to let the participants become familiarized with the correct execution of the method protocol and to optimize the LC-MS/MS conditions for mycotoxin detection, as well as to pre-check the method interlaboratory performances.

The results forms were returned by twelve out of fifteen laboratories participating in the pre-trial. One laboratory was removed because of the extremely low values reported for all target mycotoxins. A root cause analysis performed in collaboration with the participant laboratory lead us to identify a possible problem in the extract clean-up, leading to very high matrix effects which could not be properly compensated by the internal standards. Two laboratories reporting valid results were excluded from the statistical evaluations for 3-AcDON and 15-AcDON since they reported values as the sum of 3-AcDON and 15-AcDON.

The recoveries from spiked wheat flour were in the range of 78–102%, with relative standard deviations for repeatability (RSD_r) ranging from 4.4% to 12% and relative standard deviations for reproducibility (RSD_R) ranging from 8.9% to 25%. For contaminated wheat samples, RSD_r and RSD_R ranged from 3.2% to 17% and from 7.5% to 29%, respectively.

The Horwitz ratio (HorRat) values, calculated as the ratio between the RSD_R obtained during the pre-trial study and the predicted RSD_R calculated by the modified Horwitz equation [18,19], were between 0.5 and 2.1 for all mycotoxins in the three test materials. In all the considered cases, the method performances fell within the criteria established by the European Commission for the acceptability of methods for mycotoxin determination, set for DON, ZEN, and T-2/HT-2 [20,21], proving the candidate method to be suitable for the full collaborative study.

Some issues related to the method protocol execution arose and were discussed with the participants. Two participant laboratories were not able to achieve a full chromatographic separation of 3-AcDON and 15-AcDON. Two participants reporting poor peak shapes for NIV were identified as outliers by the Cochran test. Therefore, improvements of the LC conditions were suggested by the interlaboratory study coordinator and examples of suitable LC-MS/MS settings were provided (Figure 1A,B).

2.2. Full Collaborative Study

Seventeen laboratories participated in the full collaborative study, and 15 report forms were collected. Eleven participants used triple quadrupole mass detectors (in a wide range of manufacturers and models), whereas four used hybrid quadrupole-Orbitrap™ detectors. Given the limited number of participants using high-resolution MS (HRMS) detectors, low- and high-resolution MS data were evaluated as a unique set.

Three participants were not able to resolve 3- and 15-AcDON peaks and reported results relevant to the sum of the two isomers. Two participants could report data for 3-AcDON only, since 3-AcDON and 15-AcDON were detected in negative and positive ion mode, respectively, in two separate chromatographic runs, and it was not possible to quantitate 15-AcDON with the response ratio of the isotopically labelled 3-AcDON. Moreover, two participants using a triple quadrupole MS detector with polarity switching between defined retention time windows did not detect ZEN. This was probably due to slight retention time shifts leading ZEN to elute out of the dedicated retention time period.

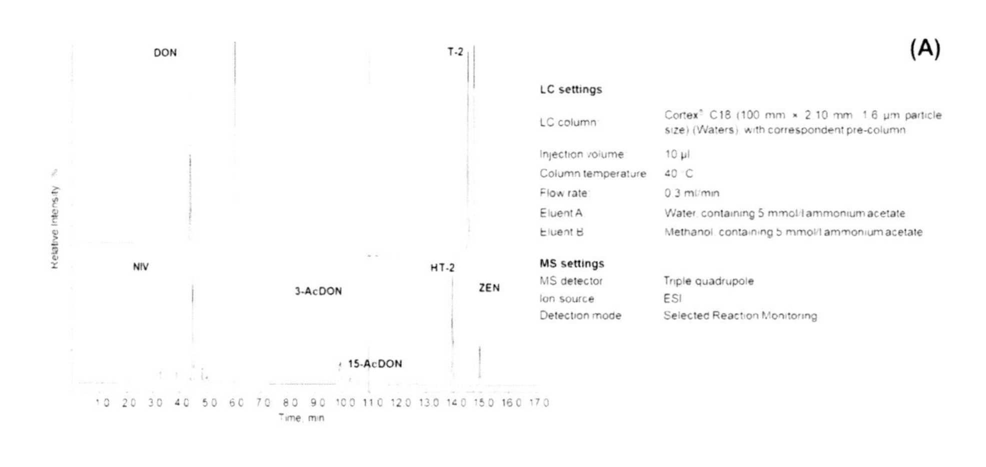

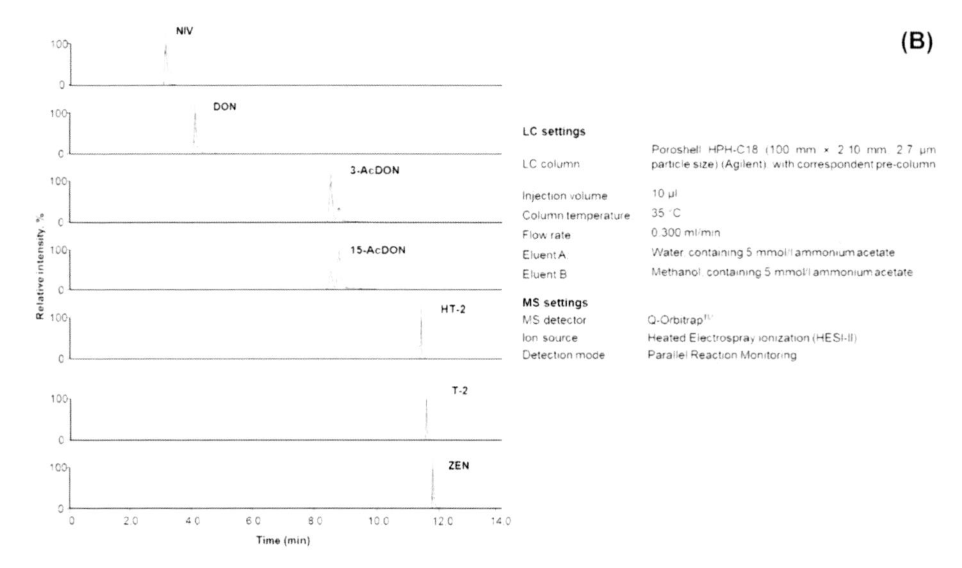

Figure 1. (**A**) Selected ion chromatogram (quantifier SRM–selected ion monitoring-transitions) and LC-MS settings of a wheat sample spiked with 120 µg/kg nivalenol (NIV), 600 µg/kg deoxynivalenol (DON), 75 µg/kg 3- and 15-AcDON (acetyl deoxynivalenol), 25 µg/kg HT-2 toxin (HT-2) and T-2 toxin (T-2), 50 µg/kg zearalenone (ZEN), and relevant isotopically labeled internal standard (ISTD—upper lines); (**B**) extracted ion chromatogram (quantifier ions) and LC-MS settings of a mycotoxin standard solution containing 1.87 µg/mL NIV, 9.37 µg/mL DON, 1.12 µg/mL 3- and 15-AcDON, 0.37 µg/mL HT-2 and T-2, 0.75 µg/mL ZEN, and relevant ISTD (upper lines).

Some data were considered "invalid" and were excluded from the statistical evaluation. Specifically, one laboratory reported problems in chromatography (poor peak shape) and provided extremely high values for the majority of the data. One laboratory using HRMS reported extremely high values in all samples for early eluting toxins (NIV and DON), probably due to poor separation from the solvent front. One participant faced sensitivity issues in all the samples but calibrants and reported extremely low values for the majority of the toxin/sample combinations. The cause was not clearly identified, but it was reasonably attributed to problems occurring during the clean-up procedure.

The results of the remaining laboratories were subjected to statistical analysis for the identification of outliers by the Cochran and Grubbs test to remove laboratories showing a significantly higher variability among replicates and extreme values, respectively [22]. For all toxin/material combinations, the number of identified outliers was lower than 20%, and thus below the maximum of 2/9 laboratories that could be removed as recommended in the AOAC (Association of Official Analytical Chemists) International guidelines for conducting interlaboratory studies [22].

Full collaborative study results are reported in Table 1. The method was validated with the perspective to be applied for compliance testing by official food control laboratories. Therefore, it was aimed at achieving method performance characteristics to meet the provisions in the Commission Regulation No. 401/2006/EC and Commission Recommendation 519/2014/EU [20,21]. The mycotoxin levels in contaminated materials were set in order to encompass legal limits (were available), which were also chosen as spiking levels for recovery assessment (Table 1).

The recoveries were in the range of 71–97%, and the lowest values (71–78%) were obtained for NIV, the most polar mycotoxin. The relative standard deviations for repeatability (RSD_r) were in the range of 2.2–34%, and the relative standard deviations for reproducibility (RSD_R) were between 6.4% and 45%. Almost all the data fulfilled the EC acceptability criteria [20,21]. In a few cases, RSD_R values higher than 40% were obtained—namely, for 3-AcDON at 75 µg/kg (wheat flour), 15-AcDON at 36 µg/kg (wheat A), 24 µg/kg (wheat flour B), 11 µg/kg (wheat crackers A), and for T-2 at 48 µg/kg (wheat flour B). Overall, the worst precision values were obtained for 15-AcDON. This could be attributed either to the lower number of valid results or the use of $^{13}C_{17}$-3-AcDON as an internal standard.

The expected HorRat values, based on historical performance data and used as a guide to determine the acceptability of the precision of a method, are between 0.5 and 1.5, although the limits for performance acceptability are 0.5–2.0 [22]. In this study, the HorRat values ranged from 0.4 to 2.0, being in some cases slightly better than the expected ones on the basis of historical data (Table 1 and Figure 2). The performances of this collaborative study are in line with the data reported for other very recent collaborative studies for the validation of multimycotoxin LC-MS/MS methods [13,23], reporting HorRat values down to 0.2. Besides the expertise of the participating laboratories, the improved performances of the LC-MS/MS multimycotoxin methods can be attributed to the use of isotopically labelled internal standards, which improved the method precision by compensating for LC-MS signal drift over long analysis sequences as well as for matrix effects.

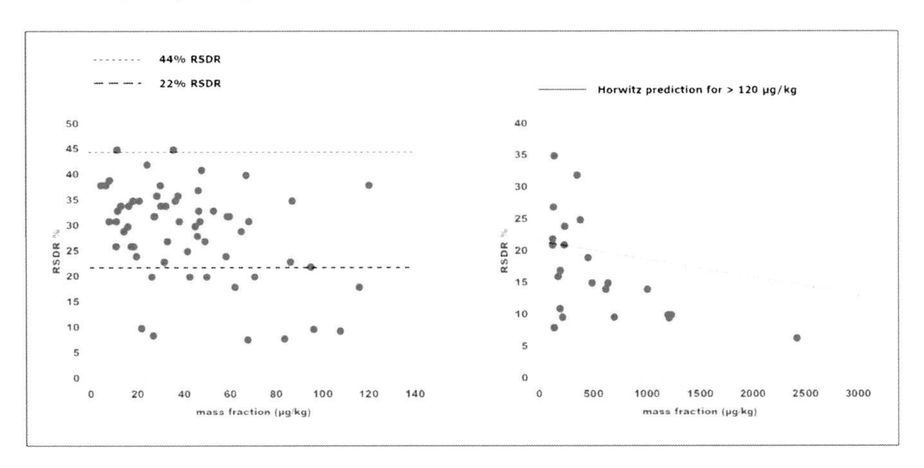

Figure 2. Reproducibility data obtained for all trichothecenes and ZEN in wheat, wheat flour, and cracker test materials in the mass fraction ranges 0–120 µg/kg and 120–2500 µg/kg.

Overall, the method validation results indicated that the focused mycotoxins can be reliably detected at levels encompassing the legal limits.

Figure 2 depicts plots of RSD_R vs. mass fraction in the range 0–120 µg/kg (a) and 120–2500 µg/kg (b). The RSD_R values were mostly between 22% and 45%.

Table 1. Interlaboratory study results for trichothecenes and zearalenone in wheat, wheat flour, and wheat crackers. Spiking levels for recovery assessment were set to be equal to the EU maximum permitted (DON, ZEN) or recommended levels (HT-2 + T-2) where available.

	Material Description	No. of Labs [a]	Mean (μg/kg)	Recovery (%)	RSD_r (%)	RSD_R(%)	HorRat [b]
NIV	Wheat, 250 μg/kg spike	14	194.1	78	5.8	17	0.8
	Wheat, contaminated A	12	87.12	n.a.	33	35	1.6
	Wheat, contaminated B	12	189.4	n.a.	11	11	0.6
	Wheat, contaminated C	13	377.8	n.a.	14	25	1.4
	Wheat flour, 150 μg/kg spike	11	116.0	77	6.1	18	0.8
	Wheat flour, contaminated A	14	52.89	n.a.	34	35	1.5
	Wheat flour, contaminated B	11	107.7	n.a.	4.2	9.4	0.4
	Wheat flour, contaminated C	11	214.8	n.a.	3.5	9.7	0.5
	Wheat crackers, 100 μg/kg spike	14	70.71	71	8.3	20	0.9
	Wheat crackers, contaminated A	12	27.69	n.a.	15	32	1.5
	Wheat crackers, contaminated B	13	64.87	n.a.	6.2	29	1.3
	Wheat crackers, contaminated C	13	123.8	n.a.	5.8	22	1.0
DON	Wheat, 1250 μg/kg spike	11	1212	97	3.7	9.5	0.6
	Wheat, contaminated A	12	635.4	n.a.	3.0	15	0.9
	Wheat, contaminated B	12	1201	n.a.	11	10	0.6
	Wheat, contaminated C	10	2420	n.a.	2.2	6.4	0.5
	Wheat flour, 750 μg/kg spike	11	692.5	92	5.0	9.7	0.6
	Wheat flour, contaminated A	14	351.0	n.a.	6.7	32	1.7
	Wheat flour, contaminated B	13	613.0	n.a.	3.8	14	0.8
	Wheat flour, contaminated C	11	1234	n.a.	3.3	10	0.7
	Wheat crackers, 500 μg/kg spike	13	448.7	90	2.8	19	1.0
	Wheat crackers, contaminated A	13	233.9	n.a.	14	24	1.2
	Wheat crackers, contaminated B	12	487.6	n.a.	5.0	15	0.8
	Wheat crackers, contaminated C	12	1005	n.a.	13	14	0.9

Table 1. *Cont.*

	Material Description	No. of Labs [a]	Mean (μg/kg)	Recovery (%)	RSD_r (%)	RSD_R(%)	HorRat [b]
3AcDON	Wheat, 150 μg/kg spike	10	136.5	91	5.7	8.0	0.4
	Wheat, contaminated A	11	49.10	n.a.	3.0	27	1.2
	Wheat, contaminated B	10	83.68	n.a.	9.4	7.9	0.4
	Wheat, contaminated C	9	67.68	n.a.	6.4	7.7	0.4
	Wheat flour, 75 μg/kg spike	11	67.01	89	8.6	40	1.8
	Wheat flour, contaminated A	11	30.22	n.a.	6.9	34	1.5
	Wheat flour, contaminated B	11	58.95	n.a.	10	32	1.4
	Wheat flour, contaminated C	9	96.22	n.a.	5.3	9.8	0.4
	Wheat crackers, 50 μg/kg spike	13	44.96	90	5.9	30	1.4
	Wheat crackers, contaminated A	12	18.47	n.a.	17	26	1.2
	Wheat crackers, contaminated B	12	26.33	n.a.	6.7	20	0.9
	Wheat crackers, contaminated C	12	49.96	n.a.	12	20	0.9
15AcDON	Wheat, 150 μg/kg spike	11	141.8	95	8.5	35	1.6
	Wheat, contaminated A	8	35.64	n.a	5.8	45	2.0
	Wheat, contaminated B	8	41.48	n.a	9.3	25	1.1
	Wheat, contaminated C	8	30.14	n.a	7.1	38	1.7
	Wheat flour, 75 μg/kg spike	9	68.25	91	6.0	31	1.4
	Wheat flour, contaminated A	8	12.87	n.a	10	34	1.5
	Wheat flour, contaminated B	9	24.38	n.a	14	42	1.9
	Wheat flour, contaminated C	9	28.64	n.a	9.6	36	1.7
	Wheat crackers, 50 μg/kg spike	12	46.24	93	8.6	37	1.7
	Wheat crackers, contaminated A	9	11.41	n.a	25	45	2.0
	Wheat crackers, contaminated B	10	16.42	n.a	10	34	1.5
	Wheat crackers, contaminated C	11	32.34	n.a	5.9	34	1.5

Table 1. *Cont.*

	Material Description	No. of Labs [a]	Mean (μg/kg)	Recovery (%)	RSD_r (%)	RSD_R(%)	HorRat [b]
HT-2	Wheat, 50 μg/kg spike	14	46.40	93	7.3	33	1.5
	Wheat, contaminated A	12	17.33	n.a.	6.6	26	1.2
	Wheat, contaminated B	14	36.29	n.a.	31	35	1.6
	Wheat, contaminated C	12	133.8	n.a.	10	27	1.2
	Wheat flour, 25 μg/kg spike	11	21.60	86	6.1	9.9	0.5
	Wheat flour, contaminated A	14	6.629	n.a.	26	38	1.7
	Wheat flour, contaminated B	13	14.18	n.a.	8.9	29	1.3
	Wheat flour, contaminated C	14	32.97	n.a.	9.2	27	1.2
	Wheat crackers, 12.5 μg/kg spike	12	10.73	86	4.7	26	1.2
	Wheat crackers, contaminated A	14	7.988	n.a.	14	31	1.4
	Wheat crackers, contaminated B	14	19.59	n.a.	15	24	1.1
	Wheat crackers, contaminated C	14	38.02	n.a.	11	31	1.4
T-2	Wheat, 50 μg/kg spike	12	45.89	92	12	28	1.3
	Wheat, contaminated A	12	27.50	n.a.	11	32	1.5
	Wheat, contaminated B	13	47.79	n.a.	34	41	1.9
	Wheat, contaminated C	12	18.01	n.a.	12	35	1.6
	Wheat flour, 25 μg/kg spike	13	20.85	83	6.3	35	1.6
	Wheat flour, contaminated A	13	11.53	n.a.	7.3	33	1.5
	Wheat flour, contaminated B	10	26.85	n.a.	6.3	8.5	0.4
	Wheat flour, contaminated C	13	37.57	n.a.	7.2	36	1.7
	Wheat crackers, 12.5 μg/kg spike	13	11.05	88	7.0	31	1.4
	Wheat crackers, contaminated A	13	4.530	n.a.	12	38	1.7
	Wheat crackers, contaminated B	14	8.091	n.a.	10	39	1.8
	Wheat crackers, contaminated C	13	15.88	n.a.	4.9	30	1.4

Table 1. *Cont.*

	Material Description	No. of Labs [a]	Mean (μg/kg)	Recovery (%)	RSD_r (%)	RSD_R(%)	HorRat [b]
ZEN	Wheat, 100 μg/kg spike	10	95.12	95	6.5	22	1.0
	Wheat, contaminated A	11	59.85	n.a.	14	32	1.5
	Wheat, contaminated B	10	125.1	n.a.	12	21	1.0
	Wheat, contaminated C	10	229.7	n.a.	11	21	1.0
	Wheat flour, 75 μg/kg spike	11	62.03	83	13	18	0.8
	Wheat flour, contaminated A	11	42.49	n.a.	13	20	0.9
	Wheat flour, contaminated B	11	86.38	n.a.	24	23	1.1
	Wheat flour, contaminated C	11	171.7	n.a.	8.3	16	0.8
	Wheat crackers, 50 μg/kg spike	12	47.13	94	7.4	31	1.4
	Wheat crackers, contaminated A	10	31.62	n.a.	17	23	1.1
	Wheat crackers, contaminated B	9	58.26	n.a.	16	24	1.1
	Wheat crackers, contaminated C	11	120.4	n.a.	7.3	38	1.7

[a] Number of laboratories (out of 15) remaining after the removal of invalid results and outliers. [b] HorRat (Horwitz Ratio) calculated using Predicted Standard Deviation from Thompson for concentrations >120 μg/kg and 22% for concentrations ≤120 μg/kg [19]. n.a.: not applicable.

Looking at the overall results, the main factors influencing recoveries and precision turned out to be the chemical nature of the toxin. Lower but still acceptable recoveries (≥71%) were obtained for NIV. With this being the most polar toxin targeted by the method, some analyte losses (higher than other mycotoxins) could occur in the solid phase extraction (SPE) clean-up loading/washing steps. Furthermore, a good chromatographic separation between 3- and 15-AcDON was confirmed to be critical to achieve satisfactory method performance results for both isomers. On the other hand, no clear influence of the matrix was observed on recoveries and precision, suggesting that the method can be considered "horizontal"—i.e., applicable to different commodities without changing/adapting the sample preparation and analysis procedure.

3. Conclusions

A method for the simultaneous determination of zearalenone and the major trichothecenes—namely, nivalenol, deoxynivalenol and its acetyl derivatives (3-acetyl DON and 15-acetyl DON), and HT2-and T-2 toxins—was validated in an international collaborative study involving 15 participant laboratories from 10 countries according to the AOAC/IUPAC (International Union of Pure and Applied Chemistry) International Harmonized Protocol [22]. The method was based on LC-MS/MS detection after SPE extract clean-up. The results of the collaborative study met the required method performance criteria, with HorRat values in the range of 0.4–2.0. Data from the collaborative study as well as feedback from participants confirmed that the chromatographic separation of 3-AcDON and 15-AcDON remains one of the challenging steps of the method, as well as achieving satisfactory recoveries for NIV, which were higher than 70%. Even though affecting the cost-effectiveness of the method, the use of isotopically labeled internal standards is essential to manage the matrix effect, contributing to achieve a satisfactory precision. Finally, the applied generic sample preparation procedure, based on polymeric SPE columns, has been already demonstrated to be applicable to a wide range of mycotoxins and food commodities [24], suggesting possible further extensions of the method scope. The method complies with the provisions of the commission regulations setting performance criteria for mycotoxin methods [20,21] and has been recently adopted as a CEN standard [17].

4. Materials and Methods

4.1. Chemicals and Reagents

Participants in the collaborative study were requested to use only reagents of recognized analytical grade and specifically the following: water, acetonitrile, and methanol of HPLC quality; ammonium acetate for mass spectrometry, $c(CH_3COONH_4) \geq 99.0\%$; solid-phase extraction columns Oasis® HLB 3 mL, 60 mg (Waters, Milan, Italy), or equivalent; standard NIV, DON, 3Ac-DON, 15Ac-DON, HT-2, T-2, ZEN crystalline, as a film or as certified standard solution; isotopically labelled internal standards ($^{13}C_{15}$-NIV, $^{13}C_{15}$-DON, $^{13}C_{17}$-3-AcDON, $^{13}C_{22}$-HT-2, $^{13}C_{24}$-T-2, $^{13}C_{18}$-ZEN) as acetonitrile solutions (25 µg/mL).

4.2. Test Materials

Three commodities—namely, common wheat (cereal), wheat flour (cereal flour), and wheat crackers (cereal-based food)—were selected. For each food commodity, one blank (containing mycotoxins at levels below the detection limit of the in-house validated method—i.e., 10 µg/kg NIV; 5 µg/kg DON; 1 µg/kg 3-AcDON and 15-AcDON; 0.5 µg/kg HT-2, T-2, and ZEN), one blank for spiking purposes (for recovery evaluation) and three contaminated materials (low, medium, and high level) with target mycotoxins were prepared. Desired mycotoxin levels in contaminated test materials were within a contamination range encompassing the EU maximum permitted levels (where applicable) (Table 1). All the materials were dispensed in plastic bottles (approx. 30 g each container) that were labeled, sealed, and stored at −20 °C until analysis for homogeneity and stability study or dispatch for collaborative trial testing.

4.2.1. Preparation of Whole Soft Wheat and Soft Wheat Flour Test Materials

Uncontaminated common wheat kernels and soft wheat flour (about 10 kg each) were selected among a set of samples purchased from the Italian retail market. The absence of contamination was verified by analysis for the mycotoxin content according to relevant validated/reference methods [25–27]. Wheat kernels were milled by an ultra-centrifugal mill (ZM 200, Retsch, 500 μm sieve). About 4 kg of each commodity were used as blank, whereas the remaining 6 kg were used for the preparation of the contaminated materials.

Since naturally contaminated materials containing all focused mycotoxins at the desired levels were unavailable at the time of this study, the preparation of contaminated test materials was performed according to a previously developed protocol [12]. Briefly, contaminated test materials were prepared by mixing and homogenizing blank material (common wheat or common wheat flour) with culture extracts of *Fusarium* toxigenic species (deposited at the Institute of Sciences of Food Production collection, http://www.ispa.cnr.it/Collection) grown on wheat kernels. Fungal cultures of *F. graminearum* ITEM 126 (producing DON, ZEN, and 3-AcDON) and *F. sporotrichioides* ITEM 707 (producing T-2 and HT-2) were dried, ground, extracted, and analyzed for the mycotoxin content according to relevant validated/reference methods [25–27]. Subsequently, a multi-mycotoxin spiking solution containing NIV, DON, 3- and 15-AcDON, T-2, HT-2, and ZEN was prepared by mixing adequate amounts of culture extracts and standard mycotoxin solutions (for NIV and 15-AcDON). For each contamination level (low, medium, high), 2 kg aliquots of blank material were split into 0.5 kg portions and fortified with the multi-mycotoxin solution. After solvent evaporation (overnight at room temperature), the 0.5 kg portions were pooled and passed through an ultra-centrifugal mill (ZM 200, Retsch, 500 μm sieve), then homogenized by a laboratory mixer for 24 h.

This procedure resulted in three contaminated wheat materials (low, medium, and high level) and three contaminated wheat flour materials (Table 1).

4.2.2. Preparation of Wheat Crackers Test Materials

Wheat crackers were prepared at the laboratory scale according to the following optimized recipe: 500 g of common wheat flour, 40 g sunflower seed, 200 mL of water, 10 g of salt. Ingredients were mixed up in a kneading machine (Princess 151 936 Silver, Milan, Italy) for 20 min. The dough was rolled out with a traditional pasta machine (model 150 Atlas Marcato SpA, Italy) into sheets of 2 mm, cut, and baked in a conventional oven at 230 °C for 10 min. Mycotoxin-contaminated wheat crackers were prepared by adding the appropriate amount of mycotoxin standard solutions to the water necessary for dough preparation. After baking, each material (blank or contaminated) was passed through an ultra-centrifugal mill (ZM 200, Retsch, 500 μm sieve), then homogenized by a laboratory mixer for 24 h.

4.2.3. Homogeneity of Test Materials

A homogeneity study was carried out according to the procedure described by ISO guide 35:2006 [28] on randomly selected units: specifically, 16 units (for blank test materials) and 11 units (for contaminated test materials) of about 25 g were taken at systematic intervals from the filling sequence. Each unit of 25 g was divided into two aliquots and analyzed in duplicate under repeatability conditions. The analytical method used for homogeneity testing was the one described in this protocol, keeping the same ratio of test portion to extraction solvent.

Homogeneity was statistically evaluated according to ISO 13528:2015 [29] and F-test. The parameters considered for the homogeneity test were the analytical precision (within bottle standard deviation, s_w–analytical SD) and the heterogeneity standard deviation (between bottle standard deviation, s_b–heterogeneity SD). The F-test was used to determine whether the observed s_b deviated significantly from the s_w.

The heterogeneity (s_b) was then compared to the target standard deviation (s). The s values were obtained using the truncated Horwitz equation corrected by Thompson—i.e., if the relative target standard deviation according to Horwitz was greater than 22%, it was truncated to 22%. The samples were considered to be adequately homogenous if $s_b \leq 0.3$ s [28,29]. Data were processed using the ProLab software (ProLab Software—QuoData, Drezden—www.quodata.de).

All the test materials passed the homogeneity test and turned out to be appropriate for the collaborative study.

4.2.4. Stability of Test Materials

Randomly selected units of the test materials were submitted to accelerated ageing at different temperatures (4 °C, 20 °C, and 60 °C) over a total period of 1.5 months according to the so-called isochronous short-term stability study [28–30]. A total of 26 bottles for each material were stored at −20 °C (reference temperature), then 2 random bottles per time were moved to the different temperatures after 0.25, 0.50, 1 and 1.5 month for a total of 24 bottles. All the units were analyzed at the end of month 1.5, under repeatability conditions, together with 2 reference samples which were kept at −20 °C over the whole period of the short-term stability study. The analytical method used for stability testing was the one described in this protocol. The stability study was performed on each contamination level per material.

Statistical results assessment was performed according to ISO guide 35:2006 using the *t*-test to test the regression for significance [28,29]. Specifically, the evaluation of data was carried out by performing a linear regression on the experimentally determined concentrations of each mycotoxin (mean values) versus time (days). For a stable material, it is expected that the intercept is equal to the reference value, whereas the slope does not differ significantly from zero.

The evaluation of data from the short-term stability study indicated that no significant trend was observed for the test samples at all temperature conditions (4 °C, 20 °C, and 60 °C) for the time span of the collaborative study. It was concluded that the three test materials were stable for at least 1.5 months following their preparation.

4.3. *Collaborative Study*

4.3.1. Study Layout

Prior to the full validation study, laboratories had to participate in a pre-trial study to let them become familiarized with the correct execution of the method protocol and optimize the LC-MS/MS conditions for mycotoxin detection. To this scope, they received:

- One blank wheat flour sample, to be used for five determinations (two determinations as blank and three determinations for recovery check).
- One wheat sample (to be analyzed as blind duplicate) contaminated with 1298 μg/kg DON; 58 μg/kg HT-2; 8.3 μg/kg T-2; 148 μg/kg ZEN.
- A mixed stock solution (stock solution B, see Section 4.3.2) and a mixed standard solution in acetonitrile to be used for spiking purposes and calibrants preparation, respectively, and a mixed isotopically labeled internal standard (ISTD) solution in acetonitrile (mixed ISTD, see Section 4.3.2) to be used as internal standard.
- Columns for solid phase extraction (SPE) clean-up.
- Method protocol in SOP (Standard Operating Procedure) format and reporting sheets.

The full validation study was planned according to the requirements of the IUPAC/AOAC international harmonized protocol [22,31]. The main purpose of the collaborative study was to estimate the precision of the candidate method under repeatability and reproducibility conditions. The method accuracy was evaluated by spiking experiments. Fifteen laboratories were involved in the trial, representing a cross-section of academia, official control, and private laboratories.

Participants received the following materials:

- Two blank materials per each commodity (wheat, wheat flour, wheat crackers), about 15 g each, to be used for two independent determinations as spiked sample for recovery checking. Participants were asked to fortify the material with the respective spiking solution with an evaporation time of approximately 1 h before the determination.
- Blind duplicates of 1 blank material and 3 contaminated materials (low, medium and high level see Table 1) per each commodity. Test material size (15 g) was sufficient to perform a single determination. Materials were coded in a random pattern.
- Three mixed mycotoxin stock solutions to spike the 3 target commodities respectively; a mixed standard solution for calibrant preparation; a mixed ISTD solution.
- Thirty (+ 2 extra) solid phase extraction (SPE) columns.
- Material receipt form.
- Standard operating protocol (SOP).
- Reporting sheets.

The results had to be expressed in micrograms per kilogram (µg/kg). Each laboratory was free to use its own LC-MS/MS set-up, and the optimization of settings for the LC and MS/MS detection was left to the participants. Mycotoxin detection was requested to be performed in Selected Reaction Monitoring (SRM) in case of MS/MS analyzers or Parallel Reaction Monitoring (PRM) in case of MS/high-resolution mass spectrometry (HRMS) analyzers. The full chromatographic separation of 3-AcDON and 15-AcDON was mandatory, and the participants were asked to provide individual data for the two toxins. The participants were requested to make available chromatograms for samples and calibration standards and to provide the following details on the applied LC-MS/MS instrumentation and method settings: LC column characteristics, mobile phase composition and gradient elution, flow rate, injection volume, MS ion source, SRM or PRM ions.

4.3.2. Mycotoxin Solutions

The following mixed mycotoxin solutions were prepared in acetonitrile, according to concentrations specified in the following:

- Mixed stock solution A, to be used for wheat spiking: NIV, 12.5 µg/mL; DON, 62.5 µg/mL; 3-AcDON, 7.5 µg/mL; 15-AcDON, 7.5 µg/mL; T-2, 2.5, µg/mL; HT-2, 2.5 µg/mL; ZEA, 5.0 µg/mL.
- Mixed stock solution B, to be used for wheat flour spiking: NIV, 7.5 µg/mL; DON, 37.5 µg/mL; 3-AcDON, 3.75 µg/mL; 15-AcDON, 3.75 µg/mL; T-2, 1.25, µg/mL; HT-2, 1.25 µg/mL; ZEA, 3.75 µg/mL.
- Mixed stock solution C, to be used for wheat crackers spiking: NIV, 5 µg/mL; DON, 25 µg/mL; 3-AcDON, 2.5 µg/mL; 15-AcDON, 2.5 µg/mL; T-2, 0.625, µg/mL; HT-2, 0.625 µg/mL; ZEA, 2.5 µg/mL.
- Mixed standard solution, prepared by 10 times dilution with acetonitrile of the multi-toxin stock solution A. This solution was used to prepare calibrants (see Table 2).
- Mixed internal standard (ISTD) solution, prepared by mixing the commercial individual ISTD solutions to obtain a mixture containing $^{13}C_{15}$-NIV, 1.25 µg/mL; $^{13}C_{15}$-DON, 6.25 µg/mL; $^{13}C_{17}$-3-AcDON, 0.75 µg/mL; $^{13}C_{22}$-HT-2, 0.25 µg/mL; $^{13}C_{24}$-T-2, 0.25 µg/mL; $^{13}C_{18}$-ZEN, 0.5 µg/mL.

Table 2. Preparation of the calibration solutions.

Mass Concentration of Calibration Solutions									
Calibration Solution	Mixed Standard Solution	Mixed ISTD Solution	NIV	DON	3-Ac DON	15-Ac DON	HT-2	T-2	ZEN
	μL	μL	μg/mL	μg/mL	μg/mL	μg/mL	μg/mL	μg/mL	μg/mL
1	25	100	0.078	0.391	0.047	0.047	0.016	0.016	0.031
2	50	100	0.156	0.781	0.094	0.094	0.031	0.031	0.063
3	100	100	0.313	1.563	0.188	0.188	0.063	0.063	0.125
4	200	100	0.625	3.125	0.375	0.375	0.125	0.125	0.250
5	400	100	1.250	6.250	0.750	0.750	0.250	0.250	0.500
6	600	100	1.875	9.375	1.125	1.125	0.375	0.375	0.750
Mass Concentration of Isotopically Labelled Analytes (μg/mL) in All Calibration Solutions									
			0.313	1.563	0.188	0.188	0.063	0.063	0.125

Aliquots of 0.5 mL of spiking solutions A, B, and C were dispensed in 2 mL amber vials. The spiking solutions were labeled as Vial #1, Vial #2, and Vial #3; mycotoxin concentrations were blind. Approximately 2 mL of mixed standard solution (labeled as Vial #4) and approximately 4 mL of mixed ISTD solution (labeled as Vial #5) were dispensed in 4 mL amber vials; the mycotoxin concentrations were specified.

To prepare the calibration solutions, different volumes of the mixed standard solution and the mixed ISTD solution were added to six autosampler vials as listed in Table 2 to obtain six calibration levels across the calibration range. After evaporation to dryness under a stream of air or nitrogen at approximately 40 °C, the dried residue was re-dissolved by adding 400 μL of HPLC injection solvent.

4.3.3. Sample Preparation

The test samples (10 g) were extracted with 50 mL (V3) acetonitrile/water 84/16 (*v*/*v*) for 60 min on an orbital shaker. The extract was filtered through filter paper (Whatman No. 4), and 5 mL (V4) of filtrate (equivalent to 1 g sample) were evaporated to dryness at 40 °C under a stream of air. The residue was reconstituted with 100 μL of methanol and then 900 μL water were added (to obtain a methanol: water ratio of 10:90, *v*/*v*). The Oasis® HLB column was activated and conditioned prior to use as follows. The column was attached to a vacuum manifold, conditioned with 2 mL methanol, and equilibrated with 2 mL methanol/water 10/90 (*v*/*v*). The reconstituted sample extract was then passed through the column at a flow rate of about one drop per second. The column was washed with 1 mL methanol/water 20/80 (*v*/*v*) and dried. Afterwards, the toxins were eluted with 1 mL methanol. To prepare the sample test solution, 100 μL of the mixed ISTD solution were added to the SPE eluate. Then the SPE eluate was evaporated to dryness under a stream of air or nitrogen at 40 °C. The dried residue was re-dissolved by adding 400 μL (V1) of HPLC injection solvent and filtered through 0.20 μm regenerated cellulose filter.

For the determination of the recoveries, spiking was performed with the mixed stock solutions A, B, and C for wheat, wheat flour, and wheat crackers, respectively, with an evaporation time of approximately one hour.

4.3.4. LC-MS Analysis

The optimization of settings for LC and MS/MS detection was left to the participants. Examples of suitable LC-MS/MS settings were provided (they are reported in Figure 1A,B). Mandatory requirements were (i) the chromatographic separation of the two isomers 15-AcDON and 3-AcDON; (ii) using a tandem mass spectrometer (MS/MS), performing Selected Reaction Monitoring (SRM) in the case of MS/MS analyzers or Parallel Reaction Monitoring (PRM) in the case of MS/high-resolution mass spectrometry (HRMS) analyzers. For mycotoxin identification, it was requested to fulfill the criteria defined in the SANTE/12089/2016 document [32].

4.3.5. Calculations

Mycotoxin quantification was performed by the isotopic dilution approach using 13C-fully labeled mycotoxins as internal standard. For each injection, the ratio of the peak area of each analyte to the peak area of the respective labelled analogue was calculated. The peak area of 15-AcDON was divided by the peak area of $^{13}C_{17}$-3-AcDON. These peak area ratios are used in all subsequent calculations. A calibration curve for each of the seven analytes (NIV, DON, 3-AcDON, 15-AcDON, HT-2, T-2, and ZEN) was prepared by plotting the peak area ratios of each analyte calculated in the calibration solutions in Table 2 (*Y*-axis, dependent variable) against the corresponding amount (μg) of analyte injected on column (*X*-axis, independent variable). The mass fraction of each mycotoxin, w, in microgram per kilogram of the sample was calculated according to Formula (1):

$$w = \left(\frac{R}{a} - \frac{b}{a}\right) \times \frac{V_1}{V_2} \times \frac{1000}{m_{SPE}} \quad (1)$$

where:

- R was the peak area ratio of the relevant analyte and the corresponding internal standard in the sample test solution;
- a was the slope of the calibration curve from calibration data, in μg^{-1};
- b was the intercept of the calibration curve from calibration data;
- V_1 was the volume of the reconstituted extract after clean-up, here: 0.4 mL;
- V_2 was the injection volume of the reconstituted sample extract, in milliliters;
- 1000 is a conversion factor;
- m_{SPE} was the sample equivalent weight purified on SPE column, here: 1 g.

The sample equivalent weight (m_{SPE}) was calculated according to Formula (2):

$$m_{SPE} = \frac{m \times V_4}{V_3} \quad (2)$$

where:

- m was the mass of the extracted test portion, here: 10 g;
- V_3 was the volume of the extraction mixture, here: 50 mL;
- V_4 was the volume of filtered extract dried before clean-up, here: 5 mL.

Author Contributions: Methodology V.M.T.L. and A.D.G.; formal analysis, B.C.; data curation, B.C., A.D.G., V.M.T.L.; writing—original draft preparation and review, V.M.T.L., A.D.G., M.P.; supervision, M.P. All authors have read and agreed to the published version of the manuscript.

Funding: The research leading to these results has received funding by the European Commission under the "Agreement for the provision of technical services to NEN", Ref.SA/CEN/ENTR/520/2013-17. Contract item: 2013-17.4.

Acknowledgments: We express our appreciation to the following collaborators for their participation in the collaborative study: Pedro A. Burdaspal, National Center for Food (Spain); Ivano Volpicella, University of Bari Aldo Moro (Italy); Sylvain Chéreau, National Institute for Agricultural Research (France); Francesca Ferrieri, Regional Agency for Environmental Protection-ARPA (Apulia, Italy); Marta Ferro, Regional Agency for Environmental Protection-ARPAL (Liguria, Italy); Carlos Goncalves and Joerg Stroka, EC-Joint Research Centre -IRMM (Belgium); Bart Huybrechts, Veterinary and Agrochemical Research Centre CODA-CERVA (Belgium); John Keegan, Public Analyst's Laboratory (Ireland); Martina Ivesic, Institute of Public Health-Andrija Stampar Teaching Institute of Public Health (Croatia); Susan MacDonald, Food and Environment Research Agency (United Kingdom); Hans Mol, RIKILT Wageningen University and Research (The Netherlands); Ivan Pecorelli, Istituto Zooprofilattico Sperimentale dell'Umbria e delle Marche (Italy); Michael Sulyok, University of Natural Resources and Applied Life Sciences-IFA Tulln (Austria); Michele Suman, Barilla G. and R. F.lli SpA (Italy); Mark W. Sumarah and Justin B. Renaud, Agriculture and Agri-Food Canada (Canada).

Conflicts of Interest: The authors declare no conflict of interest.

References

1. Miller, J.D. Mycotoxins in small grains and maize: Old problems, new challenges. *Food Addit. Contam.* **2008**, *25*, 219–230. [CrossRef] [PubMed]
2. European Food Safety Authority (EFSA). Risks to human and animal health related to the presence of deoxynivalenol and its acetylated and modified forms in food and feed. *EFSA J.* **2017**, *15*, 4718. [CrossRef]
3. European Food Safety Authority (EFSA). Panel on Contaminants in the Food Chain. Scientific Opinion on the risks for public health related to the presence of zearalenone in food. *EFSA J.* **2011**, *9*, 2197. [CrossRef]
4. European Food Safety Authority (EFSA). Panel on Contaminants in the Food Chain. Scientific Opinion on the risks for animal and public health related to the presence of T-2 and HT-2 toxin in food and feed. *EFSA J.* **2011**, *9*, 2481. [CrossRef]
5. European Food Safety Authority (EFSA). Panel on Contaminants in the Food Chain. Scientific Opinion on risks for animal and public health related to the presence of nivalenol in food and feed. *EFSA J.* **2013**, *11*, 3262. [CrossRef]
6. IARC. *Monographs on the Valuation of Carcinogenic Risks to Humans*; IARC Press: Lyon, France, 1993; Volume 56, (Suppl. 7), Available online: https://monographs.iarc.fr/list-of-classifications (accessed on 5 November 2020).
7. European Commission. Commission Regulation (EC) No 1881/2006 of 19 December 2006 setting maximum levels for certain contaminants in foodstuffs. *Off. J. Eur. Union* **2006**, *L364*, 5–24.
8. European Commission. Commission Recommendation of 27 March 2013 on the presence of T-2 and HT-2 toxin in cereals and cereal products. *Off. J. Eur. Union* **2013**, *L91*, 12–15.
9. European Food Safety Authority (EFSA). Human and animal dietary exposure to T-2 and HT-2 toxin. *EFSA J.* **2017**, *15*, 4972. [CrossRef]
10. European Food Safety Authority (EFSA). Appropriateness to set a group health based guidance value for nivalenol and its modified forms. *EFSA J.* **2017**, *15*, 4751. [CrossRef]
11. Pascale, M.; De Girolamo, A.; Lippolis, V.; Stroka, J.; Mol, G.J.H.; Lattanzio, V.M.T. Performance evaluation of LC-MS Methods for Multimycotoxin Determination. *J. AOAC Int.* **2019**, *102*, 1708–1720. [CrossRef]
12. De Girolamo, A.; Ciasca, B.; Stroka, J.; Bratinova, S.; Pascale, M.; Visconti, A.; Lattanzio, V.M.T. Performance evaluation of LC-MS/MS methods for multi-mycotoxin determination in maize and wheat by means of international Proficiency Testing. *TRAC-Trend Anal. Chem.* **2017**, *86*, 222–234. [CrossRef]
13. Bessaire, T.; Mujahid, C.; Mottier, P.; Desmarchelier, A. Multiple Mycotoxins determination in Food by LC-MS/MS: An International Collaborative Study. *Toxins* **2019**, *11*, 658. [CrossRef] [PubMed]
14. EN 16924:2017. *Foodstuffs—Determination of Zearalenone in Edible Vegetable Oils by LC-FLD or LC-MS/MS*; European Committee for Standardization: Brussels, Belgium, 2017.
15. EN 16923:2017. *Foodstuffs—Determination of T-2 Toxin and HT-2 Toxin in Cereals and Cereal Products for Infants and Young Children by LC-MS/MS after SPE Cleanup*; European Committee for Standardization: Brussels, Belgium, 2017.
16. European Union. Mandate for Standardisation Addressed to CEN for Methods of Analysis for Mycotoxins in Food. Available online: https://law.resource.org/pub/eu/mandates/m520.pdf (accessed on 5 November 2020).
17. *EN 17280:2019 Foodstuffs—Determination of Zearalenone and Trichothecenes Including Deoxynivalenol and Its Acetylated Derivatives (3-Acetyl-deoxynivalenol and 15-Acetyl-deoxynivalenol), Nivalenol T-2 Toxin and HT-2 Toxin in Cereals and Cereal Products by LC-MS/MS*; European Committee for Standardization: Brussels, Belgium, 2019.
18. Horwitz, W.; Kamps, L.R.; Boyer, K.W. Quality assurance in the analysis of foods and trace constituents. *J AOAC Int.* **1980**, *63*, 1344–1354. [CrossRef]
19. Thompson, M. Recent trends in inter-laboratory precision at ppb and sub-ppb concentrations in relation to fitness for purpose criteria in proficiency testing. *Analyst* **2000**, *125*, 385–386. [CrossRef]
20. European Commission. Commission Regulation (EC) No 401/2006 of 23 February 2006 laying down the methods of sampling and analysis for the official control of the levels of mycotoxins in foodstuffs. *Off. J. Eur. Union* **2006**, *L70*, 12–34.

21. European Commission. Commission Regulation of 16 May 2014 amending Regulation (EC) No 401/2006 as regards methods of sampling of large lots, spices and food supplements, performance criteria for T-2, HT-2 toxin and citrinin and screening methods of analysis. *Off. J. Eur. Union* **2014**, *L147*, 29–43.
22. AOAC International. Appendix D: Guidelines for Collaborative Study Procedures to Validate Characteristics of a Method of Analysis. 2005. Available online: http://www.eoma.aoac.org/appendices.asp (accessed on 16 October 2020).
23. Ye, J.; Wu, Y.; Guo, Q.; Lu, M.; Wang, S.; Xin, Y.; Xie, G.; Zhang, Y.; Mariappan, M.; Wang, S. Development and Interlaboratory Study of a Liquid Chromatography Tandem Mass Spectrometric Method for the Determination of Multiple Mycotoxins in Cereals Using Stable Isotope Dilution. *J. AOAC Int.* **2018**, *101*, 667–676. [CrossRef]
24. Lattanzio, V.M.T.; Della Gatta, S.; Suman, M.; Visconti, A. Development and in house validation of a robust and sensitive solid phase extraction: LC-MS/MS method for quantitative determination of aflatoxins B1, B2, G1, G2, ochratoxin A, deoxynivalenol, zearalenone, T-2 and HT-2 toxins in cereal-based foods. *Rapid Commun. Mass Spectrom.* **2011**, *25*, 1869–1880. [CrossRef]
25. MacDonald, S.J.; Chan, D.; Brereton, P.; Damant, A.; Wood, R. Determination of deoxynivalenol in cereals and cereal products by immunoaffinity column cleanup with liquid chromatography: Interlaboratory Study. *J. AOAC Int.* **2005**, *88*, 1197–1204. [CrossRef]
26. MacDonald, S.J.; Anderson, S.; Brereton, P.; Wood, R.; Damant, A. Determination of zearalenone in barley, maize and wheat flour, polenta, and maize-based baby food by immunoaffinity column cleanup with liquid chromatography: Interlaboratory study. *J. AOAC Int.* **2005**, *88*, 1733–1740. [CrossRef]
27. Pascale, M.; Panzarini, G.; Visconti, A. Determination of HT-2 and T-2 toxins in oats and wheat by ultra-performance liquid chromatography with photodiode array detection. *Talanta* **2012**, *89*, 231–236. [CrossRef] [PubMed]
28. International Organization for Standardization. *ISO Guide 35:2006 Reference Materials—General and Statistical Principles for Certification*; International Organization for Standardization: Geneva, Switzerland, 2006.
29. International Organization for Standardization. *ISO 13528:2015 Statistical Methods for Use in Proficiency Testing by Interlaboratory Comparison*; International Organization for Standardization: Geneva, Switzerland, 2015.
30. Lamberty, A.; Schimmel, H.; Pauwels, J. The study of the stability of reference materials by isochronus measurements. *Fresenius J. Anal. Chem.* **1998**, *360*, 359–361. [CrossRef]
31. International Union of Pure and Applied Chemistry. Protocol for the design, conduct and interpretation of collaborative studies. *Pure Appl. Chem.* **1988**, *60*, 855–864. [CrossRef]
32. SANTE/12089/2016. *Guidance Document on Identification of Mycotoxins in Food and Feed*; European Commission Directorate General for Health and Food Safety: Brussels, Belgium, 2016.

MDPI

Article

Cleaving Ergot Alkaloids by Hydrazinolysis—A Promising Approach for a Sum Parameter Screening Method

Maximilian Kuner [1], Susanne Kühn [2], Hajo Haase [3], Klas Meyer [1] and Matthias Koch [1,*]

[1] Bundesanstalt für Materialforschung und-prüfung (BAM), 12205 Berlin, Germany; maximilian.kuner@bam.de (M.K.); klas.meyer@bam.de (K.M.)
[2] Institut Kirchhoff Berlin GmbH, 13347 Berlin, Germany; susanne.kuehn@mxns.com
[3] Department of Food Chemistry and Toxicology, Technische Universität Berlin, 10623 Berlin, Germany; haase@tu-berlin.de
* Correspondence: matthias.koch@bam.de

Abstract: Ergot alkaloids are mycotoxins formed by fungi of the *Claviceps* genus, which are some of the most common contaminants of food and feed worldwide. These toxins are a structurally heterogeneous group of compounds, sharing an ergoline backbone. Six structures and their corresponding stereoisomers are typically quantified by either HPLC-FLD or HPLC-MS/MS and the values subsequently summed up to determine the total ergot alkaloid content. For the development of a screening method targeting all ergot alkaloids simultaneously, the alkaloids need to be transferred to one homogeneous structure: a lysergic acid derivative. In this study, two promising cleaving methods—acidic esterification and hydrazinolysis—are compared, using dihydroergocristine as a model compound. While the acidic esterification proved to be unsuitable, due to long reaction times and oxidation sensitivity, hydrazinolysis reached a quantitative yield in 40-60 min. Parallel workup of several samples is possible. An increasing effect on the reaction rate by the addition of ammonium iodide was demonstrated. Application of hydrazinolysis to a major ergot alkaloid mix solution showed that all ergopeptines were cleaved, but ergometrine/-inine was barely affected. Still, hydrazinolysis is a suitable tool for the development of a sum parameter screening method for ergot alkaloids in food and feed.

Keywords: ergot alkaloids; sum parameter method; hydrazinolysis; esterification

Key Contribution: Acidic esterification and hydrazinolysis were optimized and evaluated for a possible ergot alkaloid routine sum parameter screening method. Hydrazinolysis was found to be highly suitable.

Citation: Kuner, M.; Kühn, S.; Haase, H.; Meyer, K.; Koch, M. Cleaving Ergot Alkaloids by Hydrazinolysis—A Promising Approach for a Sum Parameter Screening Method. *Toxins* **2021**, *13*, 342. https://doi.org/10.3390/toxins13050342

Received: 13 April 2021
Accepted: 6 May 2021
Published: 11 May 2021

1. Introduction

Ergot alkaloids are secondary metabolites formed by Claviceps fungi of which *Claviceps purpurea* is the most common species in Europe [1,2]. The most familiar host of these fungi is rye, but wheat, triticale or other grasses may also be infested [2]. During the infestation, sclerotia—the wintering body of the fungus—are formed, which contain the toxic ergot alkaloids in varying concentrations [3]. Claviceps fungi and the corresponding alkaloids have played an important role since the Middle Ages, leading to tens of thousands of deaths [4,5]. When ingested continuously, ergot alkaloids can cause two types of disease: the gangrenous or the convulsive type of ergotism, either one eventually resulting in death [5,6].

More than 50 naturally occurring, plus further synthetic or semisynthetic, compounds belong to the group of ergot alkaloids. Six compounds and their corresponding stereoisomers (Figure 1) are most commonly found in sclerotia [7]. These are typically measured when it comes to ergot alkaloid quantification. All ergot alkaloids share a tetracyclic ring system: the ergoline structure. The major ergot alkaloids just differ in their substituents at the C8 position. Different classes can be distinguished at this substituent: clavine alkaloids, simple lysergic acid derivatives (e.g., ergometrine), ergopeptames and ergopeptines, which contain a cyclic tripeptide moiety (e.g., ergotamine, ergosine, etc.). Isomerization at the chiral C8 position leads to the formation of corresponding stereoisomers. Consequently, both isomers have to be quantified, even though the S-isomers (suffix: -inine, e.g., ergometrinine) are biologically inactive [1,7].

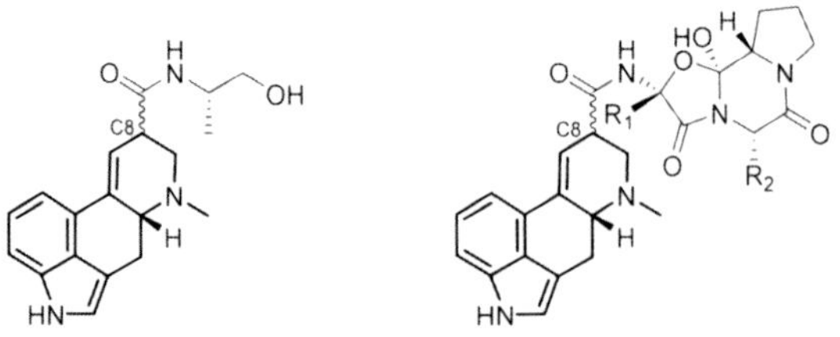

Ergometrine/-inine

R1 = -Ch3	R2 = -Ph	Ergotamine/-inine
R1 = -*i*Pr	R2 = -Ph	Ergocristine/-inine
R1 = -CH_3	R2 = -*i*Bu	Ergosine/-inine
R1 = -*i*Pr	R2 = -*i*Pr	Ergocornine/-inine
R1 = -*i*Pr	R2 = -*i*Bu	α-Ergocryptine/-inine

Figure 1. The 12 major ergot alkaloids with highlighted ergoline structure (right: ergopeptines).

For the analysis of ergot alkaloids in food/feed samples, numerous methods have been described, most of them quantifying each of the 12 major ergot alkaloids separately. Typically, HPLC with either fluorescence or MS/MS detection is used for quantification [1,8–10]. However, for the safety assessment of food samples, the fraction of each single compound to the total amount of ergot alkaloids is irrelevant, as only the sum is used for the evaluation. To date, just a few methods have been published for measuring the total amount of ergot alkaloids without quantifying each of the major ergot alkaloids individually [10–13].

The oldest method used to determine the sum of ergot alkaloids is the van Urk reaction, forming a purple compound, which is quantified colorimetrically [12]. This method is very unspecific, as the reagents used react with all indole-containing compounds in the sample, e.g., tryptophan [14]. Another quick sum parameter method is the enzyme-linked immunosorbent assay (ELISA). Due to the structural diversity of ergot alkaloids, the development of suitable antibodies to bind all ergot compounds specifically is challenging. Some ELISA kits are commercially available, but comparative studies showed poor consistency of the results obtained by ELISA when being compared to the ergot levels measured by high-performance liquid chromatography–fluorescence detection/mass spectrometric detection (HPLC-FLD/MS-MS) [15–18]. All these described sum parameter methods target the ergot alkaloids directly and thus suffer from cross-reactivity caused either by structural diversity of the ergots or an unspecific reaction (van Urk).

An approach published by Oellig et al. is to transfer the ergot alkaloids to one uniform structure [11]. Oellig et al. applied a mixture of lithium triethylborohydride and methanol to a toluene sclerotia extract. Transformation of all ergopeptines to lysergic acid amide was observed, while ergometrine/-inine remained unaffected. Subsequently, lysergic acid amide was quantified by high-performance thin layer chromatography (HPTLC-FLD). The obtained values were in good accordance with the HPLC-FLD reference data of the unreacted sample [11].

Transformation of all ergot alkaloids to one uniform compound would have several advantages. Fewer calibration standards are needed, resulting in lower costs, which is of particular interest for routine analysis. Expected shorter HPLC runtimes lead to higher throughput and thus lower the analysis costs, too. A simplification of the HPLC spectrum analysis is also expected because fewer peaks need to be integrated.

Suitable target structures are lysergic acid derivatives, as all ergot compounds share the ergoline moiety [7]. The stability of amide groups hinders the reaction, but some methods are still described in the literature attacking this bond [19]. Alkaline or acidic hydrolysis leads to the formation of either lysergic acid or lysergic acid amide, depending on the harshness of the conditions used [19–21]. The formation of the corresponding lysergic acid esters in acidic alcoholic solutions is also described in a patent [22,23]. The amide bond can also be broken with reducing agents, as described by Stoll et al. using lithium aluminum hydride ($LiAlH_4$), leading to three different products [24]. Another reaction targeting the amide bond of the ergot alkaloids is hydrazinolysis, leading to the formation of lysergic acid hydrazide [25]. Due to the good crystallization properties of the reaction product, this method can be used for the synthesis of ergot alkaloid-based drugs even with low-concentrated ergot alkaloid solutions as a starting material [26].

In brief: a sum parameter method would be highly advantageous compared to the classical HPLC-FLD or HPLC-MS/MS approach. Owing to the structural diversity of the ergot alkaloids, cleavage to a simple lysergic acid derivative as a common feature in all ergot compounds is indispensable. Some methods for the reaction of the amide bond in ergot alkaloids are described in the literature. The aim of the present study is the optimization of two ergot alkaloid cleaving methods: acidic esterification and hydrazinolysis. Furthermore, both methods should be compared not just by means of reaction yields, but also regarding other aspects for a possible routine screening method, like handling of the reaction or the possibility of parallelization. Inexpensive, easy to handle reagents and the formation of just a few defined products lead to the choice of these two reactions. The reaction conditions should be enhanced by means of reaction temperature, the addition of catalysts or microwave assistance. For the optimization experiments, dihydroergocristine (DHEC) (Figure 2) was used as a model compound, due to better availability than the native ergot alkaloids. Another advantage of DHEC is the suppression of isomerization at the C8 position during the reaction, leading to better analyzable reaction mixtures.

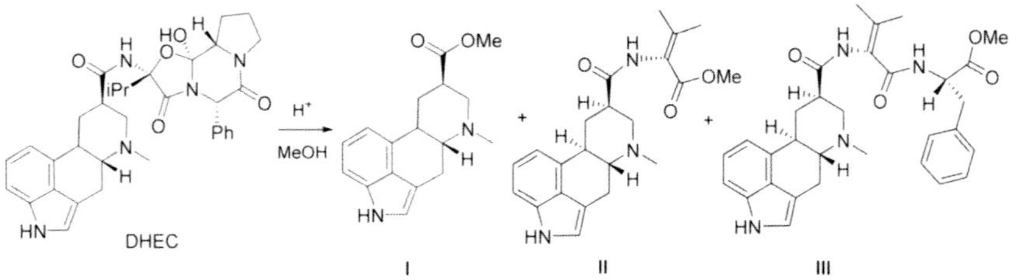

Figure 2. Acidic esterification of dihydroergocristine (DHEC). Main product: dihydrolysergic acid methylester (**I**) and main partially cleaved byproducts (**II** and **III**).

2. Results

2.1. Acidic Esterification

The acidic esterification was conducted by dissolving the corresponding acid and DHEC in methanol under nitrogen atmosphere and subsequently refluxing the solution. HPLC-MS of the reaction mixture showed, next to the desired dihydrolysergic acid methylester **I**, several partially cleaved byproducts **II** and **III** (Figure 2), whose structures were elucidated by high-resolution MS/MS experiments. Due to the side products, reaction yield determination of **I** by measuring the decrease in the DHEC concentration in the reaction mixture by HPLC-UV was impossible. NMR also proved to be unsuitable, owing to the complex reaction mixture. Hence, **I** was synthesized and purified to be used as a calibration standard in an HPLC-UV method. For the synthesis, the reaction time was prolonged to five days, so that all partially cleaved side products were transformed to **I**. The developed HPLC-UV method was used to determine the reaction yields of the acidic esterification regarding **I**, using varying reagents (Figure 3).

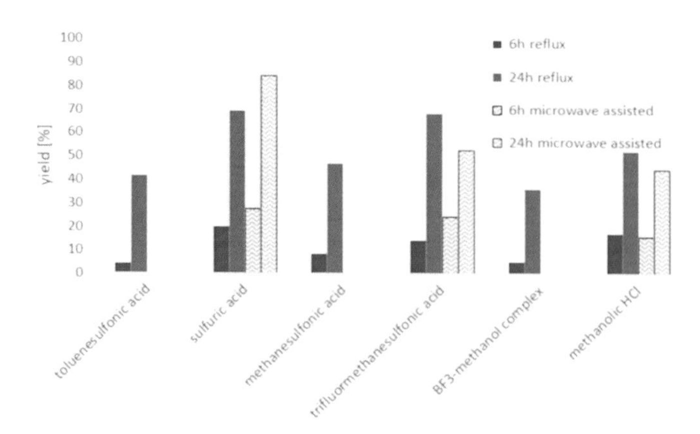

Figure 3. Yields of **I** determined by HPLC-UV (210 nm). Reaction conditions: DHEC and the corresponding acid were dissolved in methanol under inert atmosphere and refluxed (76 °C oil bath) or microwaved (set to 76 °C in temperature control mode). Samples were taken after 6 h and 24 h.

Maximum yields of about 20% were observed after 6 h and around 70% after 24 h. To increase the reaction yields, the three most promising reagents were used in a microwave-assisted approach. Compared to the previously used Schlenk flasks, keeping an inert atmosphere in the microwave reaction vessels was challenging. Thus, the heavier argon was used instead of nitrogen to prevent oxygen contamination. Despite this, the reaction had to be repeated several times, due to oxidation. The contamination of the reaction with oxygen is easily visible by an intense purple coloring of the reaction mixture. A literature search showed that ergolines can be oxidized in acidic media to a dimer showing an intense purple color at low pH [27]. An increase in the yield of **I**, when using microwave assistance, was observed for the reagents sulfuric and trifluormethane sulfonic acid after 6h. The maximum yield was 84% after 24 h using sulfuric acid.

2.2. *Hydrazinolysis*

Shimizu et al. described ammonium salts as potent accelerating reagents for the hydrazinolysis of various amides. In their study, ammonium iodide was found to be the best compound to promote the reaction [28]. Thus, the impact of ammonium iodide on the hydrazinolysis was tested. To determine the reaction yield of the hydrazinolysis, dihydrolysergic acid hydrazide **IV** was synthesized (Figure 4a) and purified. The purity of **IV** was determined by quantitative NMR (q-NMR). **IV** was used as a calibration standard and an HPLC-UV method was developed to determine the reaction yield.

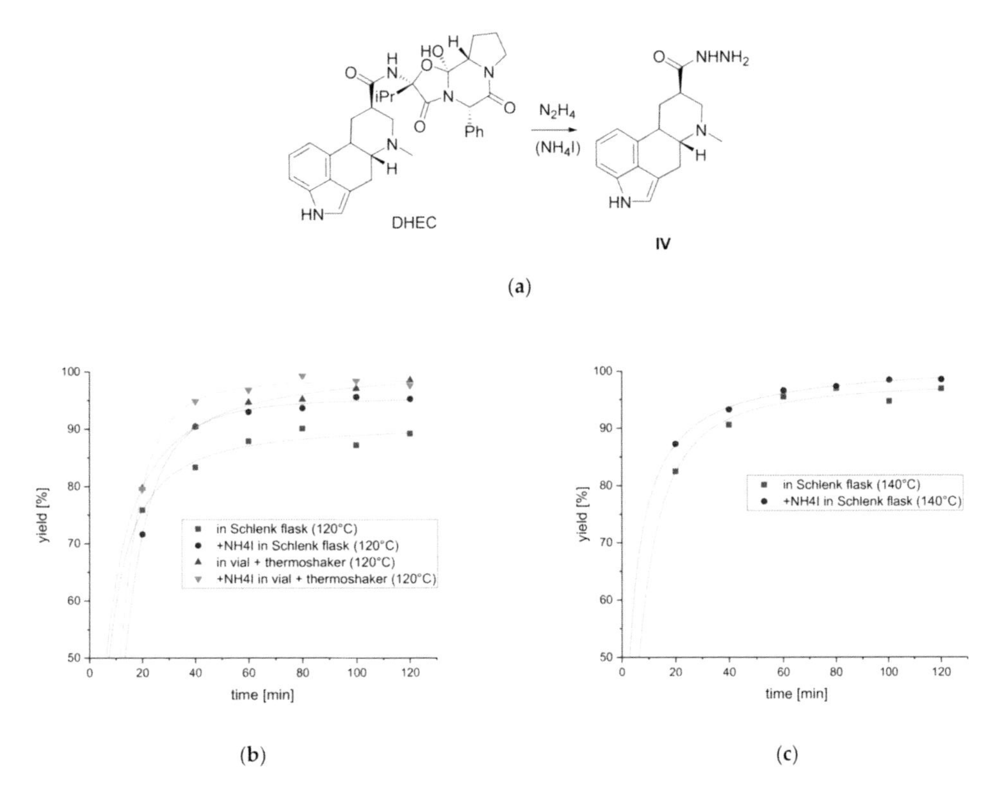

Figure 4. Hydrazinolysis of DHEC to **IV** with hydrazine hydrate. (**a**) Reaction scheme of the hydrazinolysis. To some reaction mixtures, NH_4I was added to promote the reaction as described by Shimizu et al. [28] (**b**) Yields of **IV** measured by HPLC-UV (254 nm) against reaction time at 120 °C. The reaction was conducted either in a Schlenk flask heated in an oil bath or in headspace vials heated and stirred in a thermoshaker. (**c**) Yields of **IV** measured by HPLC-UV (254 nm) against reaction time at 140 °C. The reaction was conducted in a Schlenk flask heated in an oil bath.

For the hydrazinolysis study, DHEC was suspended in hydrazine hydrate (65%) under nitrogen atmosphere. The suspension was heated, and a clear solution was obtained during the reaction. To see whether ammonium salts accelerate the reaction, one equivalent ammonium iodide was added to the reaction mixture. Additionally, different temperature levels (100 °C, 120 °C, 140 °C) were tested to optimize the reaction conditions. Samples were taken every 20 min for 2h and measured with HPLC-UV. After the first satisfactory results, the reaction vessels were changed from Schlenk flasks in an oil bath to headspace vials, which were heated and stirred in a thermoshaker. This has two advantages: the complexity of the experimental setup is reduced and thus several reactions can be performed in parallel. Headspace vials were chosen due to their good pressure resistance. The reaction yields were plotted against the reaction time (Figure 4b,c) (curves for 100 °C in the Supplementary Materials).

As expected, the yields of **IV** improved with increasing reaction temperature. Moreover, the addition of ammonium iodide improved the reaction rate, leading to an increased yield of about 5% after 40 min. Further enhancement was observed, when conducting the reaction in vials heated in a thermoshaker. Better mixing of the reaction components, when shaking the vials instead of stirring with a magnetic stirrer, led to an increased yield and shorter reaction times till quantitative reaction. Due to leakage in the septum above the boiling point of the reaction mixture, the reaction could not be properly conducted in vials at 140 °C. Therefore, these values are not included in Figure 4c.

As the results of the hydrazinolysis were quite satisfactory, hydrazinolysis was finally applied to a mix of the 12 major ergot alkaloids (see Figure 1). After 1h, all ergopeptine signals were untraceable in HPLC-FLD and HPLC-MS, while the signals of ergometrine/-inine remained nearly unchanged. Two isomeric forms of lysergic acid hydrazide were confirmed as reaction products by HPLC-MS (chromatograms in the Supplementary Materials), indicating successful cleavage of the native ergopeptines. A potential explanation for the inertness of ergometrine/-inine might be the missing neighboring effect in ergometrine, which is described in the literature for peptides with at least two vicinal amide bonds. As ergometrine contains just one amide bond, destabilization by adjacent amide moieties does not occur [29].

3. Discussion

All in all, the yields of the acidic esterification are too low after 6 h. The maximum achieved yield of 84% after 24 h would be sufficient for a screening method, but 24 h is an unacceptably long reaction time for a quick sum parameter method. With increasing pK_a (increasing strength of the acid), an increased reaction rate was observed. Enhancement could also be achieved by using microwave assistance. The biggest disadvantage of the method is the susceptibility to oxidation, which requires Schlenk flasks and nitrogen atmosphere. Thus, parallel workup of several samples is possible to a limited degree. Overall, the acidic esterification is inappropriate for use in a screening method.

The yields achieved by hydrazinolysis were satisfactory. A 95% yield of **IV** after 40 min and quantitative yield after 1 h are sufficiently quick for a sum parameter method. Adding ammonium iodide had an increasing effect on the reaction rate. Parallel cleaving of several samples is possible, as the reaction can be conducted in vials. This is also advantageous, as a further increase in the reaction yield was observed, due to the more thorough mixing of the reaction compounds in the thermoshaker. Optimum cleavage conditions for a possible sum parameter method are: 120 °C, reaction mixed and heated in a thermoshaker under inert atmosphere for 40 min in a vial. Flushing the vial with inert gas prior to hydrazinolysis is sufficient and easily practicable.

Compared to the previously published approach of Oellig et al. [11], this method is faster (2 h vs. 40 min) and involves fewer problematic reagents (hydrazine hydrate instead of lithium triethylborohydride, reacting violently with water or alcohols). Hydrazinolysis as well as the reductive approach suffer from not targeting ergometrine/-inine.

However, the hydrazinolysis cleavage reaction can be considered as a suitable approach for the development of a screening method for the ergot alkaloid content as a sum parameter. In addition, an automation of the screening method should also be feasible.

4. Materials and Methods

4.1. Chemicals and Equipment

All chemicals were used as purchased without further purification. DHEC mesylate was purchased from Teva Czech Industries s.r.o. (Prague, Czech Republic). All native ergot alkaloids were obtained from RomerLabs Division Holding GmbH (Tulln, Austria). Hydrazine hydrate, methane sulfonic acid, boron trifluoride methanol solution, tetrachloro nitrobenzene and ammonium iodide were obtained from Sigma-Aldrich (St. Louis, MO, USA). Dry methanol was purchased from Acros Organics (Ghent, Belgium). DMSO, *iso*-propanol, dichloromethane (DCM) and acetonitrile (MS grade) were obtained from Th. Geyer (Renningen, Germany). Ammonium acetate was obtained from J.T. Baker (Deventer, Netherlands). Sulfuric acid was purchased from Merck KgaA (Darmstadt, Germany). Methanolic hydrochloric acid was obtained from Bernd Kraft GmbH (Duisburg, Germany). Trifluormethane sulfonic acid was purchased from ABCR GmbH & Co KG (Karlsruhe, Germany).

Microwave-assisted reactions were conducted in an MLS 1200 Mega system (Milestone, Sorisole, Italy). For thermoshaking, an MHR-13 (HLC, Pforzheim, Germany) was used. Compounds were freeze-dried in a Gamma 1-16 LSCplus (Christ, Osterode, Germany) freeze-drying system.

HPLC-UV/-FLD/-MS were carried out with a 1290 Infinity HPLC system (Agilent, Waldbronn, Germany) coupled to a 6130 quadrupole MS (Agilent, Waldbronn, Germany). For the measurements, a Phenomenex Luna Phenyl Hexyl column (250 × 4.6 mm, 5 μm) was used.

Preparative HPLC was performed on a 1260 preparative system (Agilent, Waldbronn, Germany) coupled to a 6130 quadrupole MS (Agilent, Waldbronn, Germany). A Phenomenex Luna Phenyl Hexyl column (250 × 21.2 mm, 100 μm) was used.

High accurate masses were measured with a TripleTOF 6600 mass spectrometer (Sciex, Darmstadt, Germany) coupled to a 1290 Infinity II system (Agilent, Waldbronn, Germany). For the measurements, an Agilent Zorbax Eclipse Plus C18 column (50 × 2.1 mm, 1.8 μm) was used.

^{1}H-NMR-spectra were measured on a MercuryPlus 400 (Varian, Palo Alto, CA, USA) spectrometer at 400 MHz, ^{13}C-NMR at 100 MHz.

^{1}H quantitative NMR spectra (q-NMR) were measured on a VNMRS (Varian, Palo Alto, CA, USA) spectrometer at 500 MHz. An XP2 U/M (Mettler Toledo, Columbus, OH, USA) ultra-micro balance was used to weigh the sample and the standard (1,2,4,5-Tetrachloro-3-nitro-benzene, purity: 99.86%, *Trace*-Cert©, Sigma-Aldrich). Samples were dissolved in DMSO-d_6.

All samples and calibration curves were produced and evaluated under gravimetric control.

4.2. Synthesis of Dihydrolysergic Acid Methylester (I)

To a stirred solution of DHEC mesylate (1.522 g, 2.15 mmol) in dry methanol (50 mL), concentrated sulfuric acid (4.7 mL, 88 mmol, 41 eq) was added under nitrogen atmosphere in a Schlenk flask. The reaction mixture was heated to 55 °C and stirred for 5 days. The reaction mixture was poured into aqueous ammonia solution (50 mL 25% ammonia solution +50 mL water) and extracted with DCM (4 × 50 mL). Organic phases were collected, and all solvents removed by rotary evaporation. Clean up was conducted via preparative HPLC (acetonitrile: 0.02% aqueous NH_4Ac, 40:60, 20 min runtime). Fractions containing the target substance were combined, most of the solvent removed by rotary evaporation and the aqueous residue extracted with DCM (3 × 20 mL). The organic phase was dried with $MgSO_4$ and removed by rotary evaporation. Dihydrolysergic acid methylester (0.198 g, 0.69 mmol, 32%) was obtained as a white solid. Purity of the compound determined by q-NMR was 86.2%.

m/z (measured) (M+H)$^+$ = 285,1608 (theoretical **(M + H)$^+$**: 285,1598, δ = 3.5 ppm).

^{1}H-NMR (400 MHz, $CDCl_3$): δ (ppm) = 8.04 (s, 1H, *NH*), 7.19 (d, *J* = 1.6 Hz, 1H, CH_{ar}), 7.16 (d, *J* = 8.2 Hz, 1H, CH_{ar}), 6.95 (dt, *J* = 6.3, 1.4 Hz, 1H, CH_{ar}), 6.89 (t, *J* = 1.9 Hz, 1H, CH_{ar}), 3.76 (s, 3H, OCH_3), 3.42 (dd, *J* = 14.7, 4.3 Hz, 1H, *CH*), 3.35 – 3.26 (m, 1H, CH_2), 3.11 – 2.95 (m, 3H, *CH*), 2.75 (t, *J* = 13.0 Hz, 1H, CH_2), 2.54 (s, 3H, CH_3), 2.41 (t, *J* = 11.7 Hz, 1H, CH_2), 2.31 – 2.14 (m, 1H, CH_2), 1.71 – 1.49 (m, 1H, CH_2).

^{13}C-NMR (101 MHz, $CDCl_3$): δ (ppm) = 174.14, 133.34, 129.10, 128.55, 126.05, 123.14, 117.82, 113.29, 108.73, 66.74, 58.54, 51.78, 42.87, 41.28, 39.97, 30.49, 26.72.

4.3. Yield Determination of the Acidic Esterification

In a Schlenk flask, DHEC mesylate (78 mg, 0.11 mmol) was dissolved in dry methanol (5 mL) under nitrogen atmosphere and the corresponding acid (~55 eq) added. Due to the preset concentration of the methanolic hydrochloric acid, 200 eq of HCl were used, to keep the amount of DHEC and methanol constant compared to the other reactions. The solution was heated to 76 °C (oil bath temperature) and stirred for 24h. After 6 h and 24 h, samples were taken (~0.1 mL) and diluted with methanol (1.5 mL).

Microwave-assisted reactions were conducted using the same amounts of reagents, under argon atmosphere in microwave reaction vessels. The microwave was run in temperature control mode at 76 °C.

The obtained samples were measured with HPLC-UV (210 nm, acetonitrile: 0.02% aqueous NH_4Ac, 50:50, 0.8 mL/min, 30 min runtime). A five-point calibration (R^2: 99.7%) with concentrations of **I** in methanol ranging from 0.040 mg/g to 0.885 mg/g was prepared and used to determine the yield of **I** in the samples.

4.4. *Synthesis of Dihydrolysergic Acid Hydrazide (IV)*

Under nitrogen atmosphere in a Schlenk flask, DHEC mesylate (0.892 g, 1.26 mmol) was added to stirred hydrazine hydrate (13 mL, 263.65 mmol, 209 eq). The white suspension was heated to 140 °C for 24 h. The reaction mixture was stored in the refrigerator until a white solid precipitated. The precipitate was separated via centrifugation and the solid washed with water. Purification was conducted via preparative HPLC (acetonitrile: 0.005% aqueous NH_4Ac, 30:70, 20 min runtime). After combining all product-containing fractions, removal of the solvents by rotary evaporation and subsequent freeze-drying, the product (0.163 g, 0.57 mmol, 45%) was obtained as a white solid. Purity of the compound determined by q-NMR was 74.3%.

m/z (measured) $(M+H)^+$ = 285,1710 (theoretical **$(M+H)^+$**: 285,1710, δ = 0.0 ppm).

^{1}H-NMR (400 MHz, DMSO-d_6): δ (ppm) = 10.61 (s, 1H, *NH*), 9.10 (s, 1H, *NH*), 7.33 – 7.24 (m, 1H, CH_{ar}), 7.22 – 7.09 (m, 2H, CH_{ar}), 6.93 (dt, *J* = 7.2, 0.9 Hz, 1H, CH_{ar}), 4.67 (s, 2H, NH_2), 3.45 (dd, *J* = 14.7, 4.3 Hz, 1H, CH_2), 3.07 (ddd, *J* = 11.2, 3.8, 1.9 Hz, 1H, CH_2), 2.99 – 2.87 (m, 1H, *CH*), 2.83 – 2.60 (m, 2H, CH_2), 2.44 (s, 3H, CH_3), 2.39 (t, *J* = 11.3 Hz, 1H, CH_2), 2.15 (ddd, *J* = 11.0, 9.5, 4.3 Hz, 1H, *CH*), 2.00 (s, 1 H, CH), 1.62 (q, *J* = 12.9 Hz, 1H, CH_2).

^{13}C-NMR (101 MHz, DMSO-d_6): δ (ppm) = 172.66, 133.14, 132.36, 125.88, 122.02, 118.48, 111.92, 110.04, 108.70, 66.59, 59.29, 42.63, 40.44, 39.54, 30.73, 26.57.

4.5. *Yield Determination of the Hydrazinolysis*

In a Schlenk flask, DHEC mesylate (0.265 g, 0.378 mmol) was added to hydrazine hydrate solution (5 mL, 105 mmol, 280 eq) under nitrogen atmosphere. For the reactions with ammonium iodide, one equivalent (0.055 g, 0.378 mmol) was added. The suspension was heated to 140 °C, 120 °C or 100 °C. A clear solution was obtained during the reaction. The reaction mixture was stirred for 2h. Samples were taken every 20 min (~0.1 mL) and diluted with DMSO (~1.5 mL).

For the hydrazinolysis in vials, the same amounts were used as in the Schlenk flasks. The 20 mL headspace vials were flushed with nitrogen and sealed tightly with a septum cap after filling them with the corresponding substances. The vials were shaken in the thermoshaker for 2 h and samples taken every 20 min.

The obtained samples were measured with HPLC-UV (254 nm, acetonitrile: 0.02% aqueous NH_4Ac, 50:50, 0.8 mL/min, 25 min runtime). A five-point calibration (R^2: 99.9%) with concentrations of **IV** in DMSO ranging from 0.074 mg/g to 1.459 mg/g was used to quantify the yield of **IV**.

4.6. *Cleavage of the Native Ergot Alkaloids*

Of 1 mL ergot alkaloid standard mix solution (concentration: 0.5 ng/g for each major ergot alkaloid) in a vial, all solvents were removed to dryness in a nitrogen stream at 40 °C. After addition of 1 mL hydrazine hydrate, the vial was shaken in a thermoshaker for 1h at 120 °C. All solvents were removed in a nitrogen stream (40 °C), the sample redissolved in *iso*-propanol and measured with HPLC-FLD/-MS (HPLC conditions: acetonitrile: 0.02% aqueous NH_4Ac, 50:50, 0.8 mL/min, 60 min runtime; FLD conditions: excitation: 330 nm, emission: 415 nm; MS conditions: ESI-pos, SIM mode, $[M + H]^+$ of major ergot alkaloids and lysergic acid hydrazide).

Supplementary Materials: The following are available online at https://www.mdpi.com/article/10.3390/toxins13050342/s1: Figures S1 and S2: ^{1}H and ^{13}C-NMR spectrum of **I**. Figure S3: exemplary calibration curve for the yield determination of **I**. Figures S4 and S5: ^{1}H and ^{13}C-NMR spectrum of **IV**. Figure S6: Exemplary calibration curve for the yield determination of **IV**. Figure S7: Yield of **IV** obtained by hydrazinolysis at 100 °C. Figure S8: HPLC-FLD and HPLC-XIC of ergot alkaloid standard mixture before and after (Figure S9) hydrazinolysis.

Author Contributions: Conceptualization: M.K. (Maximilian Kuner) and M.K. (Matthias Koch); Funding acquisition: M.K. (Matthias Koch); Resources: M.K. (Matthias Koch); Investigation: M.K. (Maximilian Kuner) and K.M.; Supervision: S.K., H.H. and M.K. (Matthias Koch); Writing—original draft: M.K. (Maximilian Kuner); Writing—review and editing: S.K., H.H., K.M. and M.K. (Matthias Koch). All authors have read and agreed to the published version of the manuscript.

Funding: This research was funded by Zentrales Innovationsprogramm Mittelstand (ZIM) of the German Federal Ministry for Economic Affairs and Energy (support code: ZF4044226SB8).

Institutional Review Board Statement: Not applicable.

Informed Consent Statement: Not applicable.

Data Availability Statement: Not applicable.

Acknowledgments: The authors want to thank the ASCA GmbH for the measurement of the NMR-spectra.

Conflicts of Interest: The authors declare no conflict of interest. The funders had no role in the design of the study; in the collection, analyses, or interpretation of data; in the writing of the manuscript, or in the decision to publish the results.

References

1. Krska, R.; Crews, C. Significance, chemistry and determination of ergot alkaloids: A review. *Food Addit. Contam. Part A* **2008**, *25*, 722–731. [CrossRef]
2. Beuerle, T.; Benford, D.; Brimer, L.; Cottrill, B.; Doerge, D.; Dusemund, B.; Farmer, P.; Fürst, P.; Humpf, H.; Mulder, P.P.J. Scientific Opinion on Ergot alkaloids in food and feed. *EFSA J.* **2012**, *10*, 2798. [CrossRef]
3. Lorenz, K.; Hoseney, R.C. Ergot on cereal grains. *Crit. Rev. Food Sci. Nutr.* **1979**, *11*, 311–354. [CrossRef]
4. Hofmann, A. Historical view on ergot alkaloids. *Pharmacology* **1978**, *16* (Suppl. S1). [CrossRef]
5. Streller, S.; Roth, K. Der gehörnte Roggen. Ein chemischer Blick auf den Isenheimer Altar. *Chem. Unserer Zeit* **2009**, *43*, 272–287. [CrossRef]
6. Gabbai, L. Ergot Poisoning at Pont St. Esprit. *Br. Med. J.* **1951**, *2*, 650–651. [CrossRef] [PubMed]
7. Flieger, M.; Wurst, M.; Shelby, R. Ergot alkaloids—Sources, structures and analytical methods. *Folia Microbiol.* **1997**, *42*, 3–30. [CrossRef] [PubMed]
8. Scott, P.M. Analysis of ergot alkaloids—A review. *Mycotoxin Res.* **2007**, *23*, 113–121. [CrossRef]
9. Tittlemier, S.A.; Cramer, B.; Dall'Asta, C.; Iha, M.H.; Lattanzio, V.M.T.; Malone, R.J.; Maragos, C.; Solfrizzo, M.; Stranska-Zachariasova, M.; Stroka, J. Developments in mycotoxin analysis: An update for 2017–2018. *World Mycotoxin J.* **2019**, *12*, 3–29. [CrossRef]
10. Crews, C. Analysis of Ergot Alkaloids. *Toxins* **2015**, *7*, 2024–2050. [CrossRef] [PubMed]
11. Oellig, C. Lysergic acid amide as chemical marker for the total ergot alkaloids in rye flour—Determination by high-performance thin-layer chromatography–fluorescence detection. *J. Chromatogr. A* **2017**, *1507*, 124–131. [CrossRef]
12. Van Urk, H.W. A new sensitive reaction for the ergot alkaloids, ergotamine, ergotoxine and ergotinine and its adaptation to the examination and colorimetric determination of ergot preparations. *Pharm. Weekbl.* **1929**, *66*, 473–481.
13. Vermeulen, P.; Pierna, J.A.F.; van Egmond, H.P.; Zegers, J.; Dardenne, P.; Baeten, V. Validation and transferability study of a method based on near-infrared hyperspectral imaging for the detection and quantification of ergot bodies in cereals. *Anal. Bioanal. Chem.* **2013**, *405*, 7765–7772. [CrossRef]
14. Pindur, U. 2,2′-Diindolylmethane, 7. Mitt. Diindolylmethan-Leukobasen bei der van Urk-Reaktion mit physiologisch aktiven Indolen. *Arch. Pharm.* **1984**, *317*, 502–505. [CrossRef]
15. Kodisch, A.; Oberforster, M.; Raditschnig, A.; Rodemann, B.; Tratwal, A.; Danielewicz, J.; Korbas, M.; Schmiedchen, B.; Eifler, J.; Gordillo, A.; et al. Covariation of Ergot Severity and Alkaloid Content Measured by HPLC and One ELISA Method in Inoculated Winter Rye across Three Isolates and Three European Countries. *Toxins* **2020**, *12*, 676. [CrossRef]
16. Tunali, B.; Shelby, R.A.; Morgan-Jones, G.; Kodan, M. Endophytic fungi and ergot alkaloids in native Turkish grasses. *Phytoparasitica* **2000**, *28*, 375–377. [CrossRef]
17. Schnitzius, J.M.; Hill, N.S.; Thompson, C.S.; Craig, A.M. Semiquantitative Determination of Ergot Alkaloids in Seed, Straw, and Digesta Samples Using a Competitive Enzyme-Linked Immunosorbent Assay. *J. Vet. Diagn. Invest.* **2001**, *13*, 230–237. [CrossRef]

18. Hill, N.S.; Agee, C.S. Detection of Ergoline Alkaloids in Endophyte-Infected Tall Fescue by Immunoassay. *Crop Sci.* **1994**, *34*, 530–534. [CrossRef]
19. Komarova, E.L.; Tolkachev, O.N. The Chemistry of Peptide Ergot Alkaloids. Part 1. Classification and Chemistry of Ergot Peptides. *Pharm. Chem. J.* **2001**, *35*, 504–513. [CrossRef]
20. Stoll, A.; Petrzilka, T.; Rutschmann, J.; Hofmann, A.; Günthard, H.H. Über die Stereochemie der Lysergsäuren und der Dihydrolysergsäuren. 37. Mitteilung über Mutterkornalkaloide. *Helv. Chim. Acta* **1954**, *37*, 2039–2057. [CrossRef]
21. Stoll, A.; Hofmann, A. Zur Kenntnis des Polypeptidteils der Mutterkornalkaloide II. (partielle alkalische Hydrolyse der Mutterkornalkaloide). 20. Mitteilung über Mutterkornalkaloide. *Helv. Chim. Acta* **1950**, *33*, 1705–1711. [CrossRef]
22. Sauer, G.; Haffer, G. Process for the Preparation of Lysergic Acid Esters. U.S. Patent 4,524,208, 18 June 1985.
23. Sauer, G. Verfahren zur Herstellung von Dihydrolysergsäureestern. D.E. Patent 3,220,200 A1, 8 December 1983.
24. Stoll, A.; Hofmann, A.; Petrzilka, T. Die Konstitution der Mutterkornalkaloide. Struktur des Peptidteils. III. 24. Mitteilung über Mutterkornalkaloide. *Helv. Chim. Acta* **1951**, *34*, 1544–1576. [CrossRef]
25. Stoll, A.; Petrzilka, T.; Becker, B. Beitrag zur Kenntnis des Polypeptidteils von Mutterkornalkaloiden (Spaltung der Mutterkornalkaloide mit Hydrazin). 16. Mitteilung über Mutterkornalkaloide. *Helv. Chim. Acta* **1950**, *33*, 57–67. [CrossRef]
26. Stoll, A.; Hofmann, A. Lysergic Acid Hydrazide and a Process for Its Manufacture. U.S. Patent 2,090,429, 17 August 1937.
27. Dankházi, T.; Fekete, É.; Paál, K.; Farsang, G. Electrochemical oxidation of lysergic acid-type ergot alkaloids in acetonitrile. Part 1. Stoichiometry of the anodic oxidation electrode reaction. *Anal. Chim. Acta* **1993**, *282*, 289–296. [CrossRef]
28. Shimizu, Y.; Noshita, M.; Mukai, Y.; Morimoto, H.; Ohshima, T. Cleavage of unactivated amide bonds by ammonium salt-accelerated hydrazinolysis. *Chem. Commun.* **2014**, *50*, 12623–12625. [CrossRef]
29. Shafer, J.A.; Morawetz, H. Participation of a Neighboring Amide Group in the Decomposition of Esters and Amides of Substituted Phthalamic Acids1. *J. Org. Chem.* **1963**, *28*, 1899–1901. [CrossRef]

MDPI

Article

Simultaneous Determination of Ergot Alkaloids in Swine and Dairy Feeds Using Ultra High-Performance Liquid Chromatography-Tandem Mass Spectrometry

Saranya Poapolathep [1], Narumol Klangkaew [1], Zhaowei Zhang [2], Mario Giorgi [3], Antonio Francesco Logrieco [4] and Amnart Poapolathep [1,5,*]

1 Department of Pharmacology, Faculty of Veterinary Medicine, Kasetsart University, Bangkok 10900, Thailand; fvetsys@ku.ac.th (S.P.); fvetnak@ku.ac.th (N.K.)
2 Oil Crops Research Institute of the Chinese Academy of Agricultural Sciences, Wuhan 430062, China; zwzhang@whu.edu.cn
3 Department of Veterinary Science, University of Pisa, 56122 Pisa, Italy; mario.giorgi@unipi.it
4 Institute of Sciences of Food Production, National Research Council, 70126 Bari, Italy; Antonio.logrieco@ispa.cnr.it
5 Center of Excellence on Agricultural Biotechnology (AG-BIO/MHESI), Bangkok 10900, Thailand
* Correspondence: fvetamp@ku.ac.th; Tel.: +66-2-5797537

Abstract: Ergot alkaloids (EAs) are mycotoxins mainly produced by the fungus *Claviceps purpurea*. EAs are known to affect the nervous system and to be vasoconstrictors in humans and animals. This work presents recent advances in swine and dairy feeds regarding 11 major EAs, namely ergometrine, ergosine, ergotamine, ergocornine, ergocryptine, ergocristine, ergosinine, ergotaminine, ergocorninine, ergocryptinine, and ergocristinine. A reliable, sensitive, and accurate multiple mycotoxin method, based on extraction with a Mycosep 150 multifunctional column prior to analysis using UHPLC-MS/MS, was validated using samples of swine feed (100) and dairy feed (100) for the 11 targeted EAs. Based on the obtained validation results, this method showed good performance recovery and inter-day and intra-day precision that are in accordance with standard criteria to ensure reliable occurrence data on EA contaminants. More than 49% of the swine feed samples were contaminated with EAs, especially ergocryptine(-ine) (40%) and ergosine (-ine) and ergotamine (-ine) (37%). However, many of the 11 EAs were not detectable in any swine feed samples. In addition, there were contaminated (positive) dairy feed samples, especially for ergocryptine (-ine) (50%), ergosine (-ine) (48%), ergotamine (-ine), and ergocristine (-ine) (49%). The mycotoxin levels in the feed samples in this study almost complied with the European Union regulations.

Keywords: ergot alkaloids; swine feed; dairy feed; UHPLC-MS/MS

Key Contribution: This work describes the determination of major ergot alkaloids and their natural occurrence in swine and dairy feeds using a validated UHPLC-MS/MS with a multifunctional SPE column procedure. Our results demonstrated that the method was successfully performed according to the SANTE/11813/2017 standard. The mycotoxin levels in swine and dairy feed samples almost complied with the EU regulation.

Citation: Poapolathep, S.; Klangkaew, N.; Zhang, Z.; Giorgi, M.; Logrieco, A.F.; Poapolathep, A. Simultaneous Determination of Ergot Alkaloids in Swine and Dairy Feeds Using Ultra High-Performance Liquid Chromatography-Tandem Mass Spectrometry. *Toxins* **2021**, *13*, 724. https://doi.org/10.3390/toxins13100724

Received: 13 September 2021
Accepted: 9 October 2021
Published: 13 October 2021

1. Introduction

Mycotoxins are hazardous chemicals produced by *Aspergillus*, *Fusarium Penicillium,* and *Claviceps* genus. Mycotoxins can contaminate foods and feeds and agricultural products [1]. To date, there are more than four hundred mycotoxins with different toxicity, which have been identified in cereals, fruits, vegetables, and other agricultural commodities, resulting in potential adverse effects on human and animal health, and economic losses [2–4]. Moreover, mycotoxins are persistent in food and feeds and not completely eliminated during processing operations [3]. Recently, mycotoxins were a major category in border rejection in the European Union (EU) according to the annual report of the Rapid Alert System for Food and Feed (RASFF) [3]. The Food and Agriculture Organization (FAO) suggested one-fourth of global food crops is contaminated by mycotoxins [5]. Because of their pathogenicity and lethality, worldwide authorities including the World Health Organization (WHO) have called to monitor mycotoxins in foodstuff and feeds and set up strict maximum levels and legislation, in order to provide an early warning about mycotoxin contamination and reduce the national and international losses. In addition, the impact of climate change on *Calviceps* spp. infection of crops could result in a potential to increase the higher food safety risks for humans and animals due to mycotoxin contamination in the end products [6].

Ergot alkaloids (EAs) are toxic secondary metabolites produced by fungi of the *Claviceps* genus, mainly by the parasitic fungus *Claviceps purpurea,* which parasitize the seed heads of living plants at the time of flowering [7]. EAs are known to cause adverse health effects in humans and animals and have been found in cereals, cereal products, barley, oats, and both rye- and wheat-containing foods [8–11]. Outbreaks of ergotism in livestock do still occur, and EAs can induce abortion by its toxicity [12]. Pigs and cattle have shown symptoms after being infected with EAs, causing financial problems to both breeders and the meat industry [12,13]. Animals, including pigs exposed to EAs from grains, can cause liver and intestinal alterations [14]. In Directive 2002/32/EC on undesirable substances in animal feed and its amendment, the maximum content of rye ergot (*Claviceps purpurea*) in feed containing unground cereals has been established at 1000 mg/kg. EAs have been reported in cereals in European countries, Canada, the United States, and China [15–17]. There have also been some reports on the presence of EAs in feed from other countries, with 86–100% of EAs detected in feed samples from Germany [18] and 83% of compound feeds containing EAs with an average concentration of 89 μg/kg and a maximum concentration of 1231 μg/kg in the Netherlands [19]. The main ergot alkaloids produced by *Claviceps* species are ergometrine, ergotamine, ergosine, ergocristine, ergokryptine, and ergocornine, and the group of agroclavines [20]. Ergotamine and ergosine are heat stable whereas ergocristine, ergokryptine, ergocornine, and ergometrine are decreased by heating [21]. The conversion of ergopeptines to ergopeptinines was accelerated either by acidic or alkaline solutions. However, ergopeptinines can also be transformed to ergopeptines in organic solvents [7,22].

Studies have developed reliable analytical methods of EAs in agricultural commodities [12,15,17,19,22–26], mainly using HPLC-MS/MS. However, the challenge remains in the UHPLC-MS/MS method of optimizing the sample preparation procedure. However, signal suppression and enhancement usually occur due to the interferences in the matrix (matrix effect), leading to unreliable results [25]. To compensate for the matrix effect, some methods developed for the analysis of EAs in agricultural commodities have used a MycoSep® multifunctional column [2].

To the best of the authors' knowledge, to date, there have been a few reports on contaminations of EAs in any kinds of foodstuffs and feeds in Thailand. The current study investigated the occurrence of 11 EAs in swine and dairy feeds using a validated UHPLC-MS/MS with a multifunctional SPE column procedure. We used an SPE column for sample cleanup. Under optimization, the limit of detection, limit of quantification and linearity were studied. Accuracy and precision were evaluated as well. This work provides a promising manner to monitor EAs in feed samples.

2. Results and Discussion

2.1. Method Validation

The results of the limit of detection, limit of quantification, and linearity are reported in Table 1. From this study, the method produced good linearity.

Table 1. Performance characteristic of the analytical method: linearity ranges, limit of detection (LOD), and limit of quantification (LOQ) of the optimized LC-MS/MS method for simultaneous determination of 11 ergot alkaloids.

Ergot Alkaloid	LOD (ng/g)	LOQ (ng/g)	Calibration Range (ng/g)
Ergometrine	0.25	0.5	0.5–100
Ergosine	0.25	0.5	0.5–100
Ergocornine	0.25	0.5	0.5–100
Ergocryptine	0.25	0.5	0.5–100
Ergocristine	0.25	0.5	0.5–100
Ergotamine	0.25	0.5	0.5–100
Ergosinine	0.25	0.5	0.5–100
Ergocorninine	0.25	0.5	0.5–100
Ergocryptinine	0.25	0.5	0.5–100
Ergocristinine	0.25	0.5	0.5–100
Ergotaminine	0.25	0.5	0.5–100

Over the relevant working range, the calibration curve showed good linearity with the r^2 value higher than 0.995. The LOD value was 0.25 ng/g, and the LOQ was 0.5 ng/g (Table 1). The recovery and precision values were 70–120%, and the % relative standard deviation (RSD) values were less than 20% [27] for all 11 ergot alkaloids, as summarized in Tables 2 and 3 for the swine and dairy feeds, respectively. For identification requirements, the relative ion ratio from sample extracts was lower than 30% for all 11 ergot alkaloids [27].

Table 2. Accuracy and precision study for 11 ergot alkaloids determination in optimal LC-MS/MS conditions for swine feed samples.

Ergot Alkaloids	Spike Level, (ng/g)	Swine Feed		
		%Recovery, (%)	Intra-Day Precision, (%RSD)	Inter-Day Precision, (%RSD)
Ergometrine	0.5	113.1	2.60	8.8
	10.0	94.2	1.33	7.1
	100.0	96.2	2.95	4.7
Ergosine	0.5	111.7	5.60	8.14
	10.0	115.8	2.87	11.59
	100.0	109.2	2.88	13.38
Ergocornine	0.5	105.3	3.55	4.80
	10.0	115.1	9.39	15.69
	100.0	116.6	3.60	8.45
Ergocryptine	0.5	118.9	4.21	7.22
	10.0	109.0	11.44	11.62
	100.0	114.4	2.73	6.87
Ergocristine	0.5	107.1	8.77	9.34
	10.0	119.6	14.07	16.63
	100.0	120.0	7.45	8.95
Ergotamine	0.5	116.9	2.70	9.76
	10.0	117.3	8.68	15.67
	100.0	117.1	5.52	9.90
Ergosinine	0.5	99.1	2.25	6.51
	10.0	98.1	2.29	5.99
	100.0	97.0	1.96	5.09
Ergocorninine	0.5	101.9	4.08	8.48
	10.0	100.7	5.74	5.52
	100.0	100.0	3.07	5.84

Table 2. *Cont.*

Ergot Alkaloids	Spike Level, (ng/g)	Swine Feed		
		%Recovery, (%)	Intra-Day Precision, (%RSD)	Inter-Day Precision, (%RSD)
Ergocryptinine	0.5	110.9	2.84	5.86
	10.0	106.7	4.52	10.13
	100.0	100.2	3.52	6.11
Ergocristinine	0.5	111.3	3.96	8.71
	10.0	101.6	4.44	5.33
	100.0	98.8	1.77	6.04
Ergotaminine	0.5	100.5	3.30	7.19
	10.0	97.6	2.88	8.93
	100.0	97.0	1.64	5.69

%RSD = percentage relative standard deviation.

Table 3. Accuracy and precision study for 11 ergot alkaloids determination in optimal LC-MS/MS conditions for dairy feed samples.

Ergot Alkaloid	Spike Level, (ng/g)	Dairy Feed		
		%Recovery, (%)	Intra-Day Precision, (%RSD)	Inter-Day Precision, (%RSD)
Ergometrine	0.5	92.1	2.11	8.2
	10.0	101.1	1.40	7.2
	100.0	97.4	4.77	6.8
Ergosine	0.5	102.0	7.10	6.33
	10.0	103.0	2.22	4.55
	100.0	101.0	1.95	3.08
Ergocornine	0.5	99.6	2.98	7.78
	10.0	99.4	4.18	5.49
	100.0	92.7	4.33	10.79
Ergocryptine	0.5	101.9	2.67	6.14
	10.0	98.2	2.97	5.01
	100.0	91.5	4.10	11.83
Ergocristine	0.5	102.1	1.09	4.09
	10.0	98.1	5.51	7.02
	100.0	90.6	3.24	13.29
Ergotamine	0.5	102.4	4.13	7.85
	10.0	103.0	5.19	5.54
	100.0	101.3	1.91	2.18
Ergosinine	0.5	100.3	3.49	3.98
	10.0	100.0	1.92	3.77
	100.0	101.4	1.70	4.47
Ergocorninine	0.5	95.7	3.21	9.92
	10.0	97.7	0.97	2.32
	100.0	98.1	2.02	3.46
Ergocryptinine	0.5	97.1	2.57	8.42
	10.0	99.1	1.13	2.85
	100.0	100.4	1.82	3.34
Ergocristinine	0.5	102.7	4.51	6.73
	10.0	95.9	3.84	2.93
	100.0	98.9	1.72	3.55
Ergotaminine	0.5	101.3	1.39	4.39
	10.0	100.0	3.13	5.79
	100.0	100.3	2.05	6.73

%RSD = percentage relative standard deviation.

2.2. *Matrix Effect Study*

The study used % signal suppression/enhancement (SSE) to evaluate the matrix effects in the two types of feed matrices. If the suppression or enhancement were marginal, the %SSE would be very close to 100%; if there was strong suppression or enhancement, the %SSE would deviate from 100%. In the swine feed samples, the %SSE (94.5–106.7%) was within the acceptable range (80–120%SSE), except for ergometrine, which exhibited strong signal suppression with its %SSE (75.1%) below the acceptable range. In the dairy feed samples, the %SSE for signal suppression for the 11 ergot alkaloids was within the acceptable range 83.8–98.1%, except for ergotamine and ergometrine, which exhibited strong signal suppression (%SSE 79.6% and 44.5%, respectively). The %SSE values of the two types of feed matrices are summarized in Figure 1. For all the results of the matrix effect, the quantification of the 11 ergot alkaloids using matrix-matched calibration is necessary. The extract ion chromatograms (XIC) of spiked 11 EAs in swine and dairy feed samples were illustrated in Figures 2 and 3, respectively.

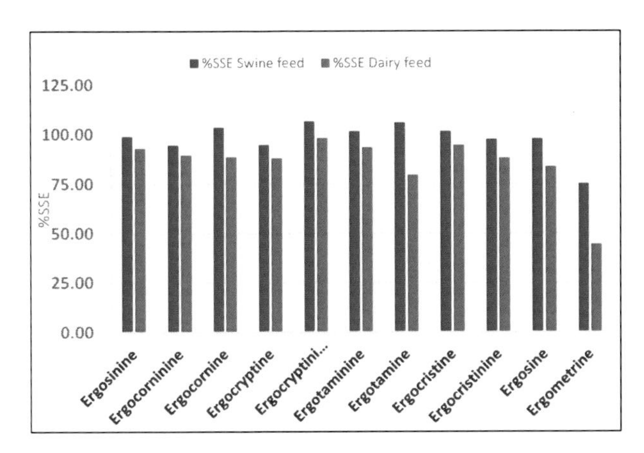

Figure 1. Signal suppression/enhancement (%SSE) for 11 ergot alkaloids in matrix-matched calibration.

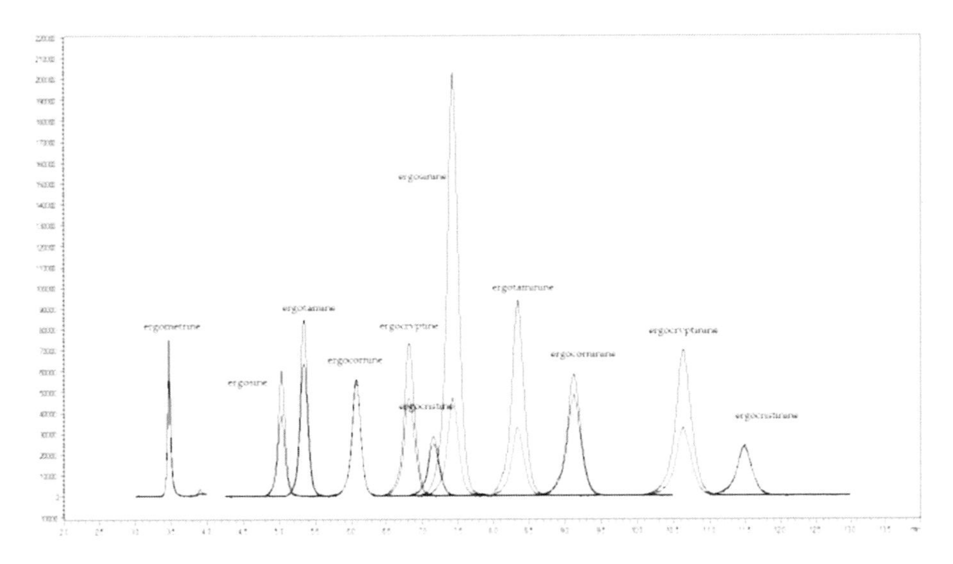

Figure 2. Extracted ion chromatogram (XIC) of spiked 11 ergot alkaloids at 20 ng/g in swine feed samples.

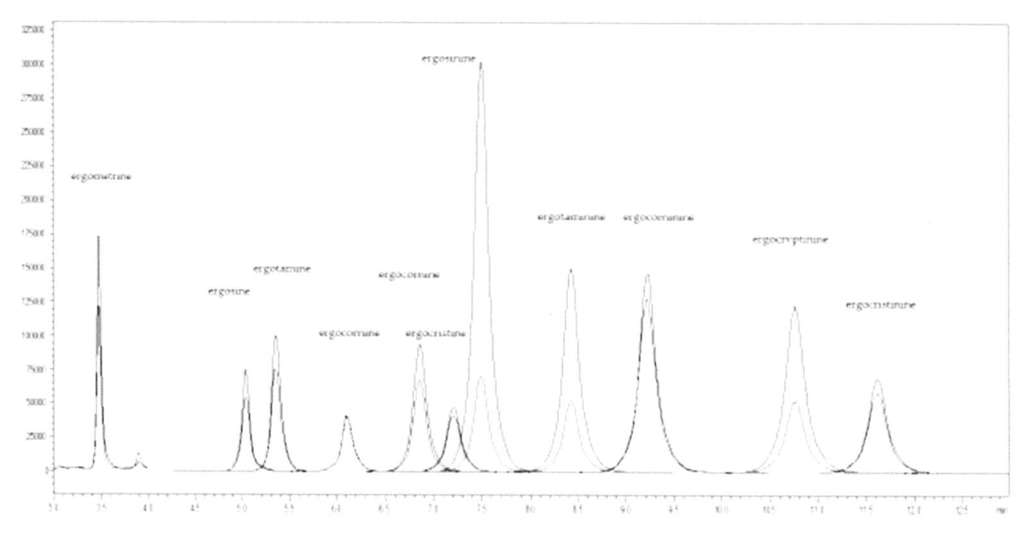

Figure 3. Extracted ion chromatogram (XIC) of spiked 11 ergot alkaloids at 20 ng/g in dairy feed samples.

2.3. Occurrence of EAs in Swine and Dairy Feeds

The method derived from this study was applied to explore the 11 ergot alkaloids in 200 feed samples consisting of swine (n = 100) and dairy feeds (n = 100). In the swine feed samples, more than 49% were contaminated with ergot alkaloids, especially ergocryptine (-ine) (40%), ergosine (-ine), and ergotamine (-ine) (37%). However, more than 50% of total samples were not detectable in 11 ergot alkaloids in the swine feed sample. The dairy feed samples had the same prevalent contaminants as the swine feed samples but with higher positive samples, especially for ergocryptine (-ine) (50%), ergosine (-ine) (48%), ergotamine (-ine), and ergocristine (-ine) (49%), as shown in Tables 4 and 5. The mycotoxin levels in all feed samples almost complied with the EU regulation ($\leq$1000 mg/kg of 11 ergot alkaloids) [28]. There are several reports on the presence of EAs in the feed from different countries, with 86–100% of listed EAs detected in feed samples from Germany [1] and 83% of compound feeds containing EAs with an average concentration of 89 μg/kg and a maximum concentration of 1231 μg/kg in the Netherlands [19]. The major detected EAs were ergosine, ergotamine, ergocristine, and ergocryptine. Interestingly, Malysheva et al. [13] reported the occurrence of EAs over three years in 1065 cereal samples originating from 13 European countries, with 52% of rye, 27% of wheat, and 44% of total samples containing EAs (ergosine, ergocristine, and ergocryptine) ranging from less than 1 to 12,340 μg/kg. In Spain, the concentrations for individual ergot alkaloids ranged between 5.9 μg/kg for ergosinine to 145.3 μg/kg for ergometrine, while the total ergot alkaloid content ranged from 5.9 to 158.7 μg/kg in swine samples. About 12.7% revealed contamination by at least one ergot alkaloid, and among contaminated swine samples, 65% were contaminated by more than one [22].

The ergot contaminations and patterns were differences due to the geographical region and environmental conditions [10].

Table 4. Occurrence of 11 ergot alkaloids in swine feed samples.

Ergot Alkaloid	Swine Feed (n = 100)		
	Number of Positive Samples	Range (ng/g)	Mean (ng/g)
Ergosinine	37	0.53–9.72	2.06
Ergosine	30	0.40–4.99	1.57
Ergocorninine	26	0.46–25.25	4.99
Ergocornine	23	0.29–4.82	1.83
Ergocryptinine	40	0.25–100.55	7.64
Ergocryptine	17	0.63–17.22	4.41
Ergotaminine	37	0.27–13.46	2.96
Ergotamine	33	0.31–18.5	3.14
Ergocristinine	28	0.67–77.6	16.15
Ergocristine	28	0.57–48.00	9.16
Ergometrine	20	0.52–10.87	3.06

Table 5. Occurrence of 11 ergot alkaloids in dairy feed samples.

Ergot Alkaloid	Dairy Feed (n = 100)		
	Number of Positive Samples	Range (ng/g)	Mean (ng/g)
Ergosinine	48	0.52–16.61	2.69
Ergosine	36	0.45–12.17	2.00
Ergocorninine	46	0.38–43.60	6.25
Ergocornine	35	0.31–11.47	2.44
Ergocryptinine	50	0.44–31.57	8.25
Ergocryptine	33	0.58–13.19	3.52
Ergotaminine	49	0.46–52.11	6.04
Ergotamine	48	0.34–43.02	5.34
Ergocristinine	49	0.62–210.53	26.75
Ergocristine	47	0.26–98.19	13.05
Ergometrine	36	0.26–31.67	2.89

3. Conclusions

EAs are hazardous mycotoxins in food and feed samples. Our results showed that the LC-ESI-MS/MS technique was an excellent tool for untargeted determination of 11 EAs in swine and dairy feed samples. The validated LC-MS/MS method using a multifunctional column was successfully performed according to the SANTE/11813/2017 standard. LODs and LOQs were recorded as 0.25 and 0.5 ng/g for EAs. Recoveries were 90.6–120%. When this technique was applied to real feed samples, it showed that 11 EAs were quantifiable in animal feeds. The mycotoxin levels in the swine and dairy samples almost complied with the EU regulations. The presence of ergot sclerotia is regulated to a maximum of 500 mg/kg in unprocessed cereal for humans [29] and 1000 mg/kg in feed materials and compound feed containing unground cereals [30]. However, further studies with a larger sample size are needed to confirm these as acceptable levels. The knowledge of toxigenic *Claviceps* species for better understanding of the production of EAs and to progress appropriate solutions for disease management should be investigated.

4. Materials and Methods

4.1. Reagents and Materials

The LC-MS/MS grade reagents, consisting of ammonium carbonate and acetonitrile (MeCN), were purchased from Fluka (St. Louis, MO, USA). The Mycosep 150 multifunctional column for extraction clean-up was purchased from Romer Labs (Tulln, Austria). Deionized water was produced using a Milli-Q system (Millipore; Bedford, MA, USA).

4.2. Analytical Standards

The analytical standards of the ergot alkaloids (ergometrine, ergosine, ergotamine, ergocornine, ergocryptine, ergocristine, ergosinine, ergotaminine, ergocorninine, ergocryptinine, and ergocristinine) were purchased from Chiron (Trondheim, Norway).

4.3. Preparation of Standards Solution

The analytical standard ergot alkaloid stock solutions were prepared in acetonitrile to provide a working standard solution of 100 μg/mL concentration for ergometrine, ergosine, ergotamine ergocryptine, ergocristine, and ergocornine and 25 μg/mL for ergosinine, ergotaminine, ergocryptinine, ergocristinine, and ergocorninine. For method validation of the spiking experiments, working standard solutions were freshly prepared at 1.0 μg/mL and were stored in amber vials at −20 °C for one week.

4.4. Sample Collection

A total of 200 feed samples consisting of swine feed (n = 100) and dairy feed (n = 100) were randomly collected from animal farms in different regions of Thailand. All samples were ground in a rotor mill ZM200 (Retsh GmbH, Hann, Germany) into a fine powder (0.50 mm) and stored at −20 °C before analysis.

4.5. Sample Preparation

The sample preparation protocol applied was developed based on Krska et al. [10]. Briefly, 5 g of homogenized feed sample was weighed into a 50 mL polypropylene (PP) centrifugation tube, followed by the addition of 25 mL of acetonitrile–ammonium carbonate buffer (3.03 mM), 84:16 (v/v). The tube was closed and shaken using a laboratory shaker (IKA Labortechnik; Staufen, Germany) for 30 min at 240 rpm. The extract was passed through Whatman No. 4 filter paper, and 4 mL of the extract was transferred to the Mycosep 150 multifunctional column (Romers lab, Tulln, Austria). Then, 1 mL of the purified extract was evaporated to dryness at 40 °C. The residue was reconstituted in 500 μL 50% mobile phase, and the mixture was passed through a 0.22 μm nylon filter before being used in the LC-MS/MS analysis.

4.6. UHPLC-MS/MS Analysis

The 11 target ergot alkaloids were analyzed using the UHPLC-MS/MS method. Chromatographic separation was developed according to Krska et al. [10]. The analysis used a Shimadzu LC-MS 8060 system (Shimadzu, Tokyo, Japan) that was equipped with a Gemini analytical column (150 × 2.0 mm i.d., 5.0 μm particle size; Phenomenex; Torrance, CA, USA) maintained at 30 °C. The mobile phase for analyses used 3.03 mM ammonium carbonate in deionized water (A) and MeCN (B) in ESI (+). The gradient elution was identical initially. The proportion of B was immediately increased from 5% to 17% within 1 min and further linearly increased to 47%, 54%, and 80% after 2, 10, and 15 min, respectively. Subsequently, the proportion of B was decreased to the initial conditions (5%) over 1 min, followed by a hold-time of 5 min, resulting in a total run-time of 21 min. The flow rate was stable at 0.5 mL/min throughout the run; 10 μL of sample extract was injected into the LC-MS/MS system.

The Shimadzu LC-MS 8060 system (Shimadzu, Japan) was equipped with an electrospray (ESI) ion source operated in positive mode. The ion source parameters were a nebulizing gas flow of 3 L/min, a heating gas flow of 10 l/min with an interface temperature: 300 °C, a CDL temperature of 250 °C, a heating block temperature of 400 °C, and a drying gas flow of 10 L/min. The dwell time (ms), Q1 Pre Bias (V), CE (V), and Q3 Pre Bias (V) were optimized during infusion of individual analytes (100 ng/mL) using automatic infusion. The MRM transitions of 11 ergot alkaloid-dependent parameters are summarized in Table 6.

Table 6. MS/MS parameters for determination of 11 ergot alkaloids.

Analyte	*m*/*z*	Dwell Time (ms)	Q1 Pre Bias (V)	CE (V)	Q3 Pre Bias (V)	Retention Time (Min)
Ergocorninine	562.40 > 223.30	70.0	−22.0	−34.0	−15.0	
	562.40 > 277.30	70.0	−26.0	−29.0	−19.0	9.4
Ergocornine	562.35 > 223.30	60.0	−22.0	−37.0	−24.0	
	562.35 > 208.20	60.0	−22.0	−45.0	−23.0	6.15
Ergocryptine	576.40 > 223.30	60.0	−22.0	−35.0	−25.0	
	576.40 > 208.30	60.0	−22.0	−49.0	−22.0	6.93
Ergocryptinine	576.35 > 223.30	80.0	−22.0	−37.0	−16.0	
	576.35 > 208.20	80.0	−22.0	−52.0	−23.0	10.99
Ergotaminine	582.30 > 223.30	70.0	−22.0	−34.0	−16.0	
	582.30 > 277.25	70.0	−22.0	−26.0	−20.0	8.57
Ergotamine	582.30 > 223.30	60.0	−22.0	−33.0	−16.0	
	582.30 > 208.20	60.0	−22.0	−44.0	−23.0	5.38
Ergocristine	610.40 > 223.30	60.0	−24.0	−36.0	−25.0	
	610.40 > 208.25	60.0	−24.0	−47.0	−22.0	7.29
Ergocristinine	610.40 > 223.30	60.0	−28.0	−36.0	−16.0	
	610.40 > 325.30	60.0	−24.0	−28.0	−22.0	11.85
Ergosine	548.45 > 223.10	60.0	−40.0	−33.0	−16.0	
	548.45 > 208.25	60.0	−40.0	−40.0	−14.0	5.05
Ergosinine	548.35 > 223.30	80.0	−20.0	−32.6	−16.0	
	548.35 > 263.10	80.0	−20.0	−27.8	−19.0	7.6
Ergometrine	326.30 > 223.30	60.0	−24.0	−25.0	−25.0	
	326.30 > 208.20	60.0	−24.0	−30.0	−22.0	3.46

4.7. Method Validation Procedure

The method performance characteristic parameters was determined to assess the efficiency of analytical method from this study by evaluating the linearity, accuracy, precision, LOD, and LOQ for EA contamination in swine and dairy feed samples. The analytes were quantified using a matrix-matched calibration standard with a pre spiking calibration curve for the 11 EAs for levels in the range 0.5–100.0 ng/g. The accuracy and precision (%RSD) were determined within the day by analyzing five replicates at three levels. The inter-day precision was determined at the same level as the within-day precision on three different days (n = 15). LODs and LOQs were calculated by analyzing the spiked samples at low level concentrations. LODs were determined as the lowest concentration of the analyte for which a signal-to-noise (S/N) ratio was 3:1, whereas S/N ratio was 10:1 for LOQs.

4.8. Matrix Effects Study

The matrix effects of the method were evaluated within two types of feed matrices: swine and dairy feed. Matrix-matched calibration curves were prepared at seven levels in the range 0.5–100.0 ng/g (n = 3 per each concentration). The matrix effects expressing the matrix-induced SSE% were defined as percentage ratios of the matrix-matched calibration slope to the solvent calibration slope. Therefore, the matrix-matched calibration curves were used for quantitative analysis.

Author Contributions: Conceptualization, A.P.; investigation, S.P., N.K. and A.P.; methodology, A.P., Z.Z. and M.G.; Resources, M.G., Z.Z. and A.F.L.; Validation, writing-original draft preparation, S.P. and A.P.; Supervision, review and editing, A.P., Z.Z., M.G. and A.F.L.; Project administration and funding acquisition, A.P. All authors have read and agreed to the published version of the manuscript.

Funding: This research is supported by the Center of Excellence on Agricultural Biotechnology, Office of the Permanent Secretary, Ministry of Higher Education, Science, Research and Innovation. (AG-BIO/MHESI).

Institutional Review Board Statement: Not applicable.

Informed Consent Statement: Not applicable.

Conflicts of Interest: The authors declare that there is no conflict of interest.

References

1. Miró-Abella, E.; Herrero, P.; Canela, N.; Arola, L.; Borrull, F.; Ras, R.; Fontanals, N. Determination of mycotoxins in plant-based beverages using QuEChERS and liquid chromatography–tandem mass spectrometry. *Food Chem.* **2017**, *229*, 366–372. [CrossRef]
2. Dzuman, Z.; Zachariasova, M.; Lacina, O.; Veprikova, Z.; Slavikova, P.; Hajslova, J. A rugged high-throughput analytical approach for the determination and quantification of multiple mycotoxins in complex feed matrices. *Talanta* **2014**, *121*, 263–272. [CrossRef]
3. Marin, S.; Ramos, A.J.; Cano-Sancho, G.; Sanchis, V. Mycotoxins: Occurrence, toxicology, and exposure assessment. *Food Chem. Toxicol.* **2013**, *60*, 218–237. [CrossRef]
4. Aupanun, S.; Poapolathep, S.; Giorgi, M.; Imsilp, K.; Poapolathep, A. An overview of toxicology and toxicokinetics of fusarenon-X, a type B trichothecene mycotoxin. *J. Vet. Med. Sci.* **2017**, *79*, 6–13. [CrossRef] [PubMed]
5. World Health Organization (WHO). Worldwide Regulation for Mycotoxins in Food and Feed in 2003. Available online: https//www.fao.org/docrep/007/y5499e/y5499e00.htm (accessed on 8 July 2021).
6. Moretti, A.; Pascale, M.; Logrieco, A.F. Mycotoxin risks under a climate change scenario in Europe. *Trends Food Sci. Tech.* **2019**, *84*, 38–40. [CrossRef]
7. Komarova, E.L.; Tolkachev, O.N. The chemistry of peptide ergot alkaloids. Part I. Classification and chemistry of ergot peptides. *Pharm. Chem. J.* **2001**, *35*, 37–45.
8. Malachova, A.; Sulyok, M.; Beltran, E.; Berthiller, F.; Krska, R. Optimization and validation of a quantitative liquid chromatography-tandem mass spectrometric method covering 295 bacterial and fungal metabolites including all regulated mycotoxins in four model food matrices. *J. Chromatogr. A* **2014**, *1362*, 145–156. [CrossRef] [PubMed]
9. Urga, K.; Debella, A.; Medihn, Y.W.; Agata, N.; Bayu, A.; Zewdie, W. Laboratory studies on the outbreak of gangrenous ergotism associated with consumption of contaminated barley in Arsi, Ethiopia. *Ethiop. J. Health Dev.* **2002**, *16*, 317–323. [CrossRef]
10. Krska, R.; Crews, C. Significance, chemistry and determination of ergot alkaloids: A review. *Food Addit. Contam. Part A* **2008**, *25*, 722–731. [CrossRef]
11. Scott, P.M. Ergot alkaloids: Extent of human and animal exposure. *World Mycotoxin J.* **2009**, *2*, 141–149. [CrossRef]
12. Craig, A.M.; Klotz, J.L.; Duringer, J.M. Cases of ergotism in livestock and associated ergot alkaloid concentrations in feed. *Front. Chem.* **2015**, *3*, 1–6. [CrossRef]
13. Malysheva, S.V.; Larionova, D.A.; Di Mavungu, J.D.; De Saeger, S. Pattern and distribution of ergot alkaloids in cereals and cereal products from European countries. *World Mycotoxin J.* **2014**, *7*, 217–230. [CrossRef]
14. Maruo, V.M.; Bracarense, A.P.; Metayer, J.P.; Vilarino, M.; Oswald, I.P.; Pinton, P. Ergot alkaloids at doses close to EU regulatory limits induce alterations of the liver and intestine. *Toxins* **2018**, *10*, 183. [CrossRef] [PubMed]
15. Tittlemier, S.A.; Drul, D.; Roscoe, M.; McKendry, T. Occurrence of ergot and ergot alkaloids in Western Canadian wheat and other cereals. *J. Agric. Food Chem.* **2015**, *63*, 6644–6650. [CrossRef] [PubMed]
16. Wyka, S.A.; Mondo, S.J.; Liu, M.; Nalam, V.; Broders, K.D. A large accessory genome, high recombination rates, and selection of secondary metabolite genes help maintain global distribution and broad host range of the fungal plant pathogen *Claviceps purpurea*. *bioRxiv* **2020**. [CrossRef]
17. Guo, Q.; Shao, B.; Du, Z.; Zhang, J. Simultaneous determination of 25 ergot alkaloids in cereal samples by ultra-performance liquid chromatography-tandem mass spectrometry. *J. Agric. Food Chem.* **2016**, *64*, 7033–7039. [CrossRef] [PubMed]
18. Ruhland, M.; Tischler, J. Determination of ergot alkaloids in feed by HPLC. *Mycotoxin Res.* **2008**, *24*, 73–79. [CrossRef] [PubMed]
19. Mulder, P.P.; Pereboom-de Fauw, D.P.; Hoogenboom, R.L.; de Stoppelaar, J.; de Nijs, M. Tropane and ergot alkaloids in grain-based products for infants and young children in the Netherlands in 2011–2014. *Food Addit. Contam. Part B* **2015**, *8*, 284–290. [CrossRef] [PubMed]
20. Kodisch, A.; Oberforster, M.; Raditschnig, A.; Rodermann, B.; Tratwal, A.; Danielewicz, J.; Kobas, M.; Schmiedchen, B.; Eifler, J.; Gordillo, A.; et al. Covariation of ergot severity and alkaloid content measured by HPLC and one Elisa method in inoculated winter rye across three isolates and three European countries. *Toxins* **2020**, *12*, 676. [CrossRef]
21. Schummer, C.; Zandonella, I.; van Nieuwenhuyse, A.; Moris, G. Epimerization of ergot alkaloids in feed. *Heliyon* **2020**, *6*, e04336. [CrossRef]
22. Arroyo-Manzanares, N.; Rodriguez-Estevez, V.; Garcia-Campana, A.M.; Castellon-Rendon, E.; Gamiz-Gracia, L. Determination of principal ergot alkaloids in swine feeding. *J. Sci. Food Agric.* **2021**, *101*, 5214–5224. [CrossRef] [PubMed]
23. Di Mavungu, J.D.; Malysheva, S.A.; Sanders, M.; Larionova, D.; Robbens, J.; Dubruel, P.; van Peteghem, C.; de Saeger, S. Development and validation of a new LC-MS/MS method for the simultaneous determination of six major ergot alkaloids and their corresponding epimers. Application to some food and feed commodities. *Food Chem.* **2012**, *135*, 292–303. [CrossRef]

24. Versilovskis, A.; Mulder, P.J.; Pereboom-de Fauw, P.K.H.; de Stoppelaar, J.; de Nijs, M. Simultaneous quantification of ergot and tropane alkaloids in bread in the Netherlands by LC-MS/MS. *Food Addit. Contam. Part B* **2020**, *13*, 215–223. [CrossRef] [PubMed]
25. Malysheva, S.V.; di Mavungu, J.D.; Goryacheva, I.Y.; de Saeger, S. A systematic assessment of the variability of matrix effects in LC-MS/MS analysis of ergot alkaloids in cereals and evaluation of method robustness. *Anal. Bioanal. Chem.* **2013**, *405*, 5595–5604. [CrossRef] [PubMed]
26. Babic, J.; Tavcar-Kalcher, G.; Celar, F.A.; Kos, K.; Cervek, M.; Jackovac-Strajn, B. Ergot and ergot alkaloids in cereal grains intended for animal feeding collected in Slovenia: Occurrence, pattern and correlations. *Toxins* **2020**, *12*, 730. [CrossRef] [PubMed]
27. SANTE, J. European Commission Document No. SANTE/11813/. Guidance Document on Analytical Quality Control and Method Validation Procedures for Pesticides Residues Analysis in Food and Feed. 2017, pp. 3357–3367. Available online: https//ec.europa.eu/food/sites/food/files/plant/docs/pesticides_mrl_guidelines_wrkdoc_2017-11813.pdf (accessed on 9 July 2021).
28. European Commission. Commission Regulation (EC) 472/2002 of March 12th 2002 amending Regulation (EC) 466/2001 setting maximum levels for certain contaminants if foodstuffs. *Off. J. Eur. Communities* **2002**, *L75*, 18–20.
29. European Commission. Commission Regulation (EC) No 1881/2006 of 19 December 2006 setting maximum levels for certain contaminants in foodstuffs. *Off. J. Eur. Union* **2006**, *L364*, 5–24.
30. European Commission. Directive 2002/32/EC of the European Parliament and of the Council of 7 May 2002 on undesirable substances in animal feed. *Off. J. Eur. Communities* **2002**, *L140*, 10–21.

Article

Validation of New ELISA Technique for Detection of Aflatoxin B1 Contamination in Food Products versus HPLC and VICAM

Elsayed Hafez [1], Nourhan M. Abd El-Aziz [2], Amira M. G. Darwish [2], Mohamed G. Shehata [2,*], Amira A. Ibrahim [1], Asmaa M. Elframawy [3] and Ahmed N. Badr [4]

1 Department of Plant Protection and Biomolecular Diagnosis, Arid Lands Cultivation Research Institute (ALCRI), City of Scientific Research and Technological Applications (SRTA-City), Alexandria 21934, Egypt; elsayed_hafez@yahoo.com (E.H.); amiranasreldeen@yahoo.com (A.A.I.)
2 Department of Food Technology, Arid Lands Cultivation Research Institute (ALCRI), City of Scientific Research and Technological Applications (SRTA-City), Alexandria 21934, Egypt; nourhanm.abdo@gmail.com (N.M.A.E.-A.); amiragdarwish@yahoo.com (A.M.G.D.)
3 Nucleic Acids Research Department, Genetic Engineering & Biotechnology Research Institute (GEBRI), City of Scientific Research and Technological Applications (SRTA-City), Alexandria 21934, Egypt; asmaameg@yahoo.com
4 Food Toxicology and Contaminants Department, National Research Centre, Dokki, Cairo 12622, Egypt; noohbadr@gmail.com
* Correspondence: gamalsng@gmail.com

Citation: Hafez, E.; Abd El-Aziz, N.M.; Darwish, A.M.G.; Shehata, M.G.; Ibrahim, A.A.; Elframawy, A.M.; Badr, A.N. Validation of New ELISA Technique for Detection of Aflatoxin B1 Contamination in Food Products versus HPLC and VICAM. *Toxins* **2021**, *13*, 747. https://doi.org/10.3390/toxins13110747

Received: 16 July 2021
Accepted: 8 October 2021
Published: 21 October 2021

Publisher's Note: MDPI stays neutral with regard to jurisdictional claims in published maps and institutional affiliations.

Abstract: AbstractToxin-contaminated foods and beverages are a major source of illness, may cause death, and have a significant negative economic impact worldwide. Aflatoxin B1 (AFB1) is a potent toxin that may induce cancer after chronic low-level exposure. This study developed a quantitative recombinant *AflR* gene antiserum ELISA technique for aflatoxin B1 detection in contaminated food products. Aflatoxin B1 residuals from 36 food samples were analyzed with HPLC and VICAM. DNA was extracted from aflatoxin-contaminated samples and the *AflR* gene amplified using PCR. PCR products were purified and ligated into the pGEM-T vector. Recombinant plasmids were sequenced and transformed into competent *E. coli* (BL21). Molecular size and B-cell epitope prediction for the recombinant protein were assessed. The purified protein was used to induce the production of IgG antibodies in rabbits. Serum IgG was purified and labeled with alkaline phosphatase. Finally, indirect-ELISA was used to test the effectiveness of polyclonal antibodies for detection of aflatoxin B1 in food samples.

Keywords: aflatoxin B1; recombinant *AflR* gene; VICAM; HPLC; I-ELISA; peanut; wheat flour; milk powder

Key Contribution: The indirect-ELISA technique was compared with HPLC and the VICAM system in food samples. The new technique showed reasonable accuracy; cost- and time-effective detection of aflatoxin B1.

1. Introduction

Mycotoxins are toxic secondary metabolites produced naturally by many fungi under certain growth conditions. These toxins affect metabolic processes and cause disease and death in humans and animals [1]. Toxicological actions of mycotoxins are recognized but few of these compounds or their derivatives are identified as antibiotics, growth promoters, or other drugs. Major mycotoxins include aflatoxins, gliotoxin, citrinin, ergot alkaloids, fumonisins, ochratoxin, and patulin [2].

Humans do not produce antibodies to mycotoxins and cannot be immunized against their toxicity. Nearly 25% of food becomes inedible due to contamination with mycotoxins; aflatoxins are the most serious source of contamination [3]. In 1993, aflatoxin is classified as a class one carcinogen by the World Health Organization Cancer Research Institute.

Aflatoxin causes hepatotoxicity in both humans and animals. Exposure to this toxin can lead to liver cancer and death. The chemical is a bifuran toxoid produced by strains of *Aspergillus flavus* and *Aspergillus parasiticus*. About 20 derivatives are recognized, for example, B1, B2, G1, G2, M1, and M2. Aflatoxin B1 (AFB1) is the most potent and carcinogenic [4]. Aflatoxins M1 and M2 are hydroxylated metabolites of aflatoxin B1 produced by animals and commonly exist in milk and dairy products. The toxins are not common in grains. AFB1 is detected on the surfaces of maize and peanuts.

Toxicity of and exposure to AFB1 has been extensively investigated. The toxin may result in severe disease, including carcinogenesis, mutagenesis, growth retardation, and immune suppression [5]. The aflatoxin-producing fungus, *P. flavus*, grows and produces aflatoxins on preharvest maize and on maize in storage [6]. Peanuts are also susceptible to *Aspergillus* infection in the field or during storage. Both maize and peanuts are rich nutrient sources for these fungi [7]. Aflatoxin-contaminated agricultural products may pose serious health risks to humans and animals and negatively affect international trade [8]. According to the Food and Drug Administration (FDA) in the USA, an acceptable aflatoxin level in food is 0 ppb.

Typically, high-performance liquid chromatography (HPLC) and liquid chromatography mass spectrometry (LC-MS) are used for quantification of aflatoxins. LC-MS/MS can detect trace levels, but some limits exist. Chemical detection is slower than spectral detection (hours vs. seconds) and tedious. Professional analysts and precise chemical instruments are required [9]. Thus, accurate, rapid, full-scale detection of AFB1 is important in assessing human health and economic impacts. Evaluating contaminated food directly for specific fungi using genes that control aflatoxin is a promising strategy [10]. Gallo et al. [11] reported such genes in the genome of aflatoxin-producing fungi. However, the authors' method required costly instruments, amplification, isolation, and quantification along with trained personnel [12]. The method is quite complex and costly for routine use. This study aimed to develop a new cost- and time-effective quantitative method using modified recombinant *AflR* gene antiserum enzyme-linked immunoassay (ELISA) for aflatoxin B1 detection in contaminated food products.

2. Results and Discussion

2.1. The Aflatoxin B1 Detection with HPLC and VICAM

VICAM was less sensitive than HPLC analysis for aflatoxin AFB1 detection in several samples—peanut 2, flours 2 and 3, and milk-powder 3. HPLC is widely used for the analysis of aflatoxins for sensitivity and accuracy [13] (Table 1). HPLC is an excellent quantitative method in detection of aflatoxins [14], although it requires skilled operators, extensive sample preparation, and is a high-cost equipment [15].

Table 1. HPLC and VICAM screening analysis of 36 food samples for aflatoxin detection.

Samples	HPLC	VICAM
Peanut 1	+++	+++
Peanut 2	+++	++
Peanut 3	+++	+++
Flour 1	++	++
Flour 2	+++	++
Flour 3	++	+
Milk-powder 1	++	++
Milk-powder 2	+	+
Milk-powder 3	++	+

+: refers to low aflatoxin contamination level; ++: refers to moderate aflatoxin contamination level; +++: refers to high aflatoxin contamination level.

2.2. Molecular Detection and SDS-PAGE

A unique band at about 760 bp was observed in all positive samples (Figure 1A). Negative amplification was observed in flour contaminated with aflatoxin (50, 75, and 100 mg). Amplified DNA was cloned and in vitro transcribed protein was separated on SDS-PAGE (Figure 1B). SDS-PAGE analysis revealed a protein of about 28 kDa. Molecular weight determination regarded as first characterization step of protein was further used in toxin detection.

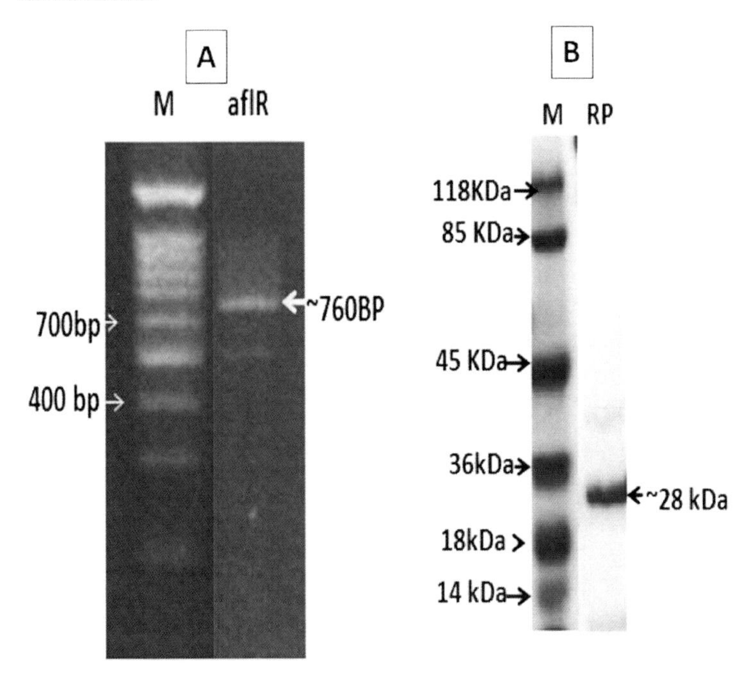

Figure 1. (**A**) PCR product amplified using the specific primers of the aflR gene (Aflatoxin B1). M: DNA marker and aflR the amplified gene in molecular size about 760 bp. (**B**) The recombinant protein of the in vitro transcribed aflR gene (Aflatoxin B1) with molecular size about 28 kDa.

2.3. Sequence Analysis and Phylogenic Construction

The PCR product was purified and sequenced and a 750 bp fragment was obtained. The sequence was aligned using the NCBI analysis tool and showed 97% similarity with other *AflR* genes listed in GeneBank. The sequence was compared with 50 sequences and a phylogenetic tree was constructed. An Egyptian *AflR* gene was closely similar to *AflR* gene MH752587 obtained from *Aspergillus* sp. PS-2018c isolate BN038G *AFlR*, Arizona, USA (Figure 2).

2.4. Antigenicity Test

A 256 residue amino acid sequence was deduced from the gene sequence (Table 2). Eight peptides showed antigenic activity by Kolaskar & Tongaonkar Antigenicity [16] Figure 3. Peptide lengths ranged from 8 to 14 amino acids. Their positions started from amino acid numbers 26, 66, 107, 136, 170, 186, 205, and 236 (Table 2). Epitope prediction using Kolaskar and Tongaonkar Antigenicity Prediction identifies the protein epitopes that are useful for diagnostic purposes and also in the development of peptide vaccines [17].

Figure 2. Phylogenetic tree for the amplified aflatoxin B1 based on the DNA nucleotide sequence and compared with the other 50 *AFB1*genes listed on gene bank. The phylogeny was constructed using Mega 6 program.

Table 2. Predicted peptides with antigenic activity, their length, and positions.

No.	Start	End	Peptide	Length	>aflIR d Deduced Amino Acid Sequence
1	26	33	LMQVPKIY	8	MSHSYNTFAGWFINTPTGRTQGSLA LMQVPKI
2	66	76	EHYLLFLVQFV	11	YLAGNKSFLGSQPAHDGLRYLEPEACMRAGQL
3	107	120	TPQLVTFVYIHLDL	14	AEHYLLFLVQFV NRSRSSLVTRFQPRYVNKEC
4	136	143	FTLCVPPRLA	8	TARQSLGQVRT PQLVTFVYIHLDL SARQRKGO
5	170	179	PGRCVPPPRLA	10	ATLQEKAF TLCVFFPA FNSKLYSTPSSRPPRW
6	186	196	IAVRVVPVQKC	11	LTIFPPGHI PGRCVPPRLA ALESSG IAVRVVPVQKC
7	205	215	VLGVSNVVLPV	11	DAPRRNRP VLGVSNVVLPV NTWSPSGWAAT
8	227	236	RALPVPLIQL	10	RALPVPLIQL GDHQRVFLQPDRNRDIRRIT

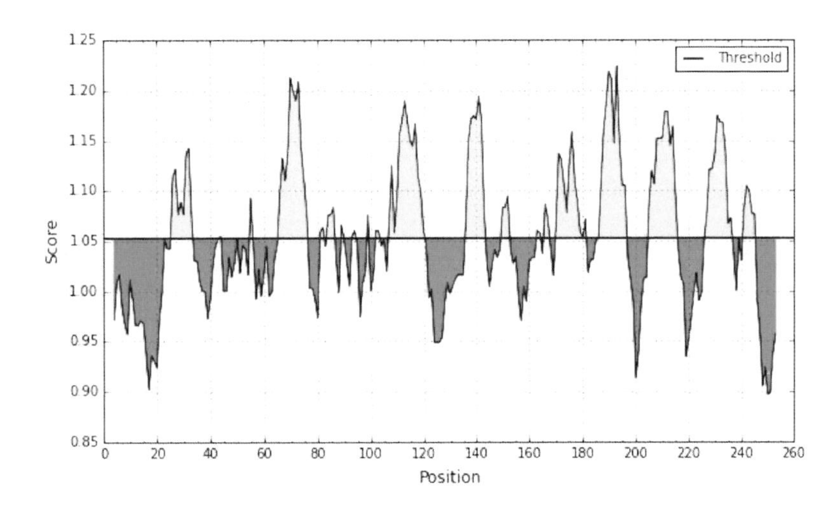

Figure 3. The predicted antigenic activity of the recombinant protein (aflR).

2.5. IgG Polyclonal Antibody Labeling and Purification

Serum obtained from immunized rabbits was fractionated using affinity chromatography protein G-Sepharose column and one band of conventional IgG with a molecular weight of 130 kDa was obtained. Moreover, two bands of a 42 kDa heavy chain and a 19 kDa light chain were separated under reducing conditions. Glutaraldehyde was used to prepare conjugates using a ratio of 4:1 of IgG and enzyme alkaline phosphatase (AP). IgG-AP conjugates were purified by gel filtration on a Sephacryl S200 column. AP (EC 3.1.3.1) is a stable enzyme its activity can be measured by many different substrates. The most common method of labeling immunoglobulin G (IgG) antibody with this enzyme uses the homobifunctional reagent glutaraldehyde [18].

2.6. Validation of the Modified Recombinant AflR Gene Antiserum ELISA Technique with HPLC and VICAM

In definition, validation is establishing the performance specifications of a new diagnostic tool such as a new test, laboratory developed test or modified method. But verification is defined as one-time process to determine performance characteristics of a test before use in patient testing [19].

ELISA was unable to distinguish among antigens due to the presence of common epitopes on protein surfaces [20–22]. Sampling/sub-sampling variation significantly affects the accuracy of aflatoxin analysis [23]. Extracts of 36 samples were used for validation to minimize the impacts of such variation.

Recombinant antiserum detected *AflR* recombinant protein within a concentration range 0–1000 pg/mL with a linear correlation between *AflR* antigenic protein and absorbance at 405 nm ($y = 0.0014x - 0.0148$; $R^2 = 0.9946$) (Figure 4). Non-significant differences among three samples of the same product, peanut, flour, or milk powder, were observed after HPLC ($p > 0.05$) (Table 3). The VICAM method showed similar results. However, the modified ELISA showed significant differences among toxin detections in these product samples. The serum-based analysis confirmed specific PCR results. No false negatives were observed in I-ELISA results and false positives were either nil or negligible.

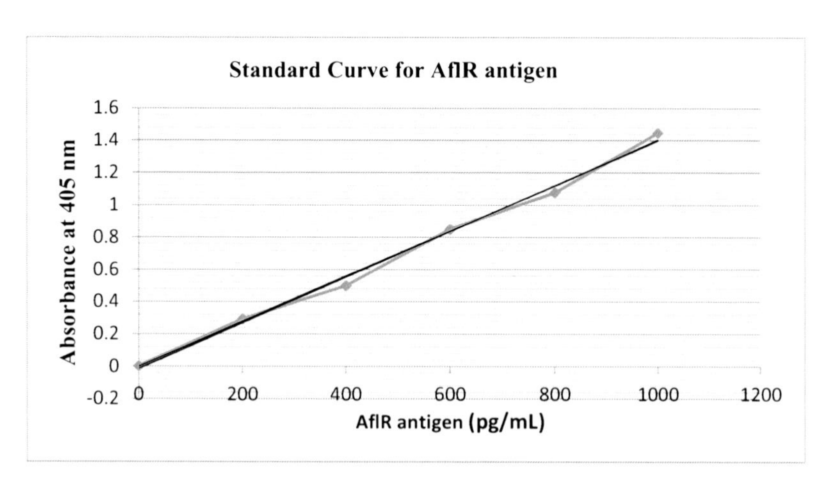

Figure 4. I-ELISA standard curve for *AflR* recombinant protein using purified serum IgG.

Table 3. Comparative results obtained by HPLC, VICAM, specific PCR, and I-ELISA (ng/g).

Sample	HPLC (ng/g)	VICAM (ng/g)	Specific PCR	ELISA (ng/mL)
Peanut 1	3.26 ± 0.68 [a]	5.86 ± 0.58 [a]	+	3.43 ± 0.40 [a,b]
Peanut 2	2.83 ± 0.58 [a]	5.46 ± 1.26 [a]	+	4.76 ± 0.92 [a]
Peanut 3	2.50 ± 0.10 [a]	6.20 ± 0.45 [a]	+	2.36 ± 0.90 [b,c]
Flour 1	0.60 ± 0.23 [b]	1.90 ± 0.36 [c]	+	1.98 ± 0.94 [b,c]
Flour 2	0.44 ± 0.22 [b]	1.83 ± 0.35 [c]	+	3.33 ± 0.51 [a,b]
Flour 3	0.66 ± 0.15 [b]	2.20 ± 0.20 [c]	+	1.82 ± 1.01 [c]
Milk-powder 1	0.93 ± 0.71 [b]	2.90 ± 0.70 [b,c]	+	4.26 ± 0.81 [a]
Milk-powder 2	0.82 ± 0.50 [b]	3.83 ± 0.55 [b]	+	1.55 ± 0.67 [c]
Milk-powder 3	1.23 ± 0.62 [b]	3.60 ± 0.91 [b]	+	1.46 ± 0.84 [c]

The mean values indicated in the same column within variable with different superscripts (a, b, and c) were significantly different ($p < 0.05$); +: present of fungal infection.

Although the correlation between the data in Figure 5A,B (comparing HPLC against VICAM and ELISA) reflected that the correlation of HPLC against VICAM (Figure 5A) was better than ELISA. On the other hand, a good correlation was observed between ELISA and VICAM (Figure 5C). However, the represented modified ELISA technique is easier to use, economic as it does not need sophisticated chemicals or highly trained technicians, have a good sensitivity to detect low infection levels determining aflatoxin B1 in foods and can represent a successful alternative in case other approaches are hard to be reached in less developed communities. Previous observations were reported for validation of a competitive direct SUNQuik ELISA for aflatoxin in peanuts using a reference HPLC method and other methods, including a minicolumn and the VICAM Afla test system [24]. The comparison between HPLC, VICAM, and validated method I-ELISA with respect to limit of detection, precision and accuracy, time of analysis, cost of analysis, and use of organic solvents is summarized in Table 4.

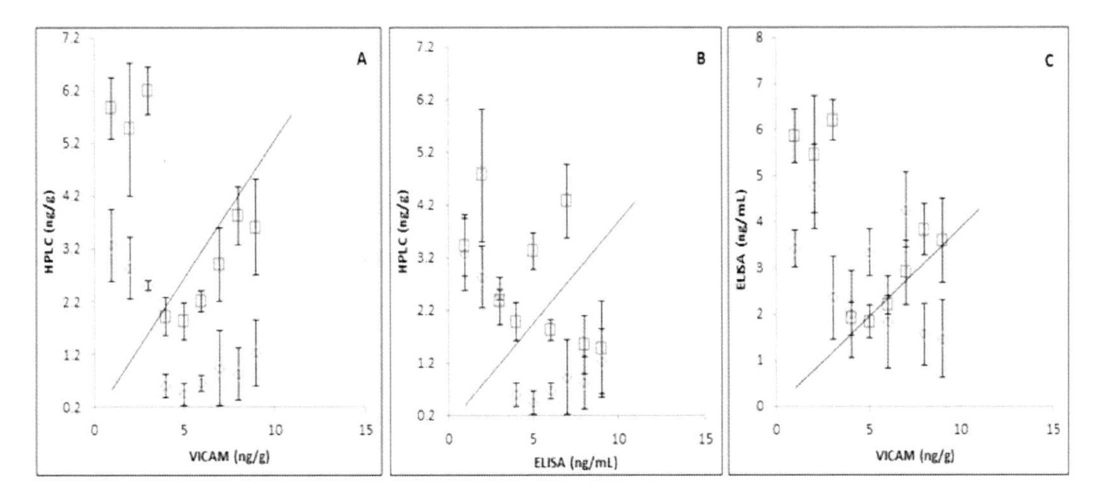

Figure 5. Correlations between HPLC and VICAM (**A**), HPLC and ELISA (**B**), ELISA and VICAM (**C**).

Table 4. Comparison between HPLC, VICAM, and validated method I-ELISA.

Parameters	HPLC	VICAM	I-ELISA
Limit of detection	<0.008 ng/mL	<2 ng/mL	<1 ng/mL
Time of analysis	120 min	90 min	30 min
Cost of analysis	High	Moderate	Moderate
Use of organic solvents	Yes	Yes	No

2.7. Limitations of the Modified Recombinant AflR Gene Antiserum ELISA Technique

Our new established method has many limitations that must be clarified to determine and specify the application field of this method. First, our new ELISA technique does not measure the aflatoxins itself, hence, this type of test cannot be used for official control. However, it could be useful for auto control and rapid results and decision-making within a farm/company. Second, although this method is quantitative test, all positive results need to be confirmed with a confirmatory method such as HPLC or LC-MS as it is based on the measurement of a recombinant protein controlled by the gene responsible for aflatoxin biosynthesis, but not on the toxin itself.

3. Conclusions and Future Perspective

Aflatoxin B1 detection is an increasingly important health and economic issue. Accurate detection is essential to assess health problems in both humans and animals. Conventional detection methods are time-consuming and require expensive chemicals and apparatus (HPLC and VICAM). An accurate and rapid detection method that requires fewer chemicals is needed. We developed a specific quantitative detection technique (I-ELISA) using recombinant *AflR* protein. *AflR* is involved in aflatoxin biosynthesis. Comparison of results achieved from the new modified ELISA with other standardized methods HPLC and VICAM, revealed that the new ELISA technique can be used at many applications as an economic alternative to detect low levels of aflatoxin contamination. This modified technique may address problems associated with the reliable and rapid detection of aflatoxin B1 contamination in food products. The technique could be used to develop highly sensitive (0–1000 pg/mL) testing capabilities. In future, hybridoma cell culture antibody production technique can be used for production of antibodies against AflR protein for large-scale manufacturing of rapid I-ELISA kit. This method will yield a production scale ranging from milligram to gram level with competitive pricing.

4. Materials and Methods

4.1. Sampling

Thirty-six food samples of three food products (12 samples each) were collected from a local market in Alexandria, Egypt. Products were prepared by different companies (4 packages each). Samples were peanuts (300 g packages), wheat flour (2 kg packages), and milk powder (500 g packages). Aflatoxin B1 was extracted for subsequent analysis.

4.2. HPLC Detection

One mL of each sample was centrifuged at 6000 rpm for 15 min, then filtered through a 0.45 μm hydrophobic polytetrafluoroethylene syringe filter in preparation for gel pores chromatographic (GPC) analysis. The supernatant was transferred to 1.5 mL micro-tubes and passed through an immune-affinity column at a rate of 1–2 drops/s. The column was washed twice with 10 mL water: methanol (90:10) at a flow rate of 3 mL/min. Aflatoxins were eluted by slowly passing 1 mL of methanol through the column. The clear eluent was then repassed through a 0.45 μm filter [25]. Subsequently, 100 μL of trifluoracetic acid and 200 μL n-hexane were added to samples and mixed by vortexing for 30 s. The vial was left for 15 min before addition of 900 μL of water: acetonitrile, 9:1 and remixing by vortexing. The hexane layer was then removed and samples were analyzed for AFs as previously reported [26] using a Waters HPLC system, Model 6000, a solvent delivery system, and a Model 720 system controller equipped with a fluorescence detector (Model 274) at excitation and emission wavelengths of 360 and 450 nm, respectively. Separation used 5 μm of sample, a Waters symmetry column (150 × 4.6 mm id), and a flow rate of 1 mL/min with an isocratic system of 1% acetic acid: methanol: acetonitrile (55:35:10).

4.3. AflA-Vt Detection

Afla-V strip tests utilize the proven sensitivity and selectivity of VICAM monoclonal antibodies to accurately detect and measure aflatoxins B1 at levels of 2 to 100 ppb. These samples were subjected to aflatoxin extraction and quantification using the VICAM fluorometry method. Briefly, representative samples (100 g) of shelled peanuts were added with 10 g of NaCl and 200 mL of methanol/water (80:20 *v*/*v*), homogenized using a Waring blender at high speed for 1 min and filtered through Whatman paper. Five ml of the filtrate was diluted with 20 mL HPLC water then re-filtered. Ten milliliter filtrate was purified with VICAM immunoaffinity columns (VICAM Aflatest, MA, USA) containing aflatoxin-specific (B1) monoclonal antibodies and washed with 10 mL HPLC water before the aflatoxin was eluted with 1 mL methanol. The eluted fraction was diluted twice with HPLC water and measured with the VICAM fluorometer (VICAM Series 4EX Fluorometer). All procedures were done according to the manufacturer's instructions [27].

4.4. Specific PCR Detection Method

DNA from food samples was extracted using a QiaGene DNA extraction kit (QiaGene, Berlin, Germany). DNA was dissolved in DEPC-treated water, quantified spectrophotometrically and analyzed using 1.2% agarose gels. The AflR gene (744 bp) was amplified using specific primers. The PCR reaction consisted of 1 μL of DNA in 2.5 μL Taq polymerase buffer 10× (Promega, New York, NY, USA) containing a final concentration of 1 mM $MgCl_2$, 0.2 Mm dNTPs, 20 pmol of each primer, and 0.2 μL Taq polymerase (5 U/μL) in a final reaction volume of 25 μL. The PCR reaction program was: initial denaturation at 95 °C for two minutes followed by 35 cycles of 58 °C for one min, 72 °C for one min, and 95 °C for 2 min. A final extension step at 72 °C for 5 min was included at the end of the reactions. PCR amplification products were separated in 1.5% agarose with 0.5× TBE buffer and visually analyzed with a gel documentation system (Syngene) [28]. Forward (5'-TAAGCAGAATTCGAATAGCTTCGCAGGGTGGT-'3) and reverse (5'-GAATAGCTTCGCAGGGTGGTGCGGCCGCTAAGCA-'3) primers were designed by Primer-Blast, NCBI.

4.5. Detection via AflR Gene Analysis and Transformation

4.5.1. Cloning, Sequencing, and *AflR* Gene Transformation

The PCR product (Section 2.4) was excised from the gel and purified using a QIA quick gel extraction kit (Qiagen Inc., Berlin, Germany). Purified DNA was ligated into the pGEM-T vector (Promega Co., New York, NY, USA). Recombinant plasmids were directly sequenced using an automated sequencer (Macrogene Company, Seoul, Korea), with a universal vector primer. DNA homology searches were carried out using the NCB1 databases and the BLAST network service. *EcoR*I and *Not*I restriction enzymes were used for gene release and insertion into the pPROEX HTa expression vector (Life Technologies, New York, NY, USA). The recombinant plasmid was transformed into competent *E. coli* (BL21) cells and recombinant protein was recovered as previously described [29].

4.5.2. Molecular Size Determination of *AflR* Recombinant Protein

The recombinant protein was separated on 12% SDS PAGE and molecular size determined using a standard low range protein marker (BioRad, Hercules, CA, USA). Gel preparation, staining, and destaining were carried out following Laemmli [30].

4.5.3. Epitope Prediction and Antigenic Determination

B-cell epitope prediction analysis was performed following Kolaskar and Tongaonkar [16] to examine the epitope in different antigenic determinants.

4.5.4. Immunization and Antibody Production

Rabbit Immunization with *AflR* Recombinant Protein

Ten male New Zealand White rabbits, age 10–16 weeks and weighing 3.5–4.0 kg were used. Physical examinations confirmed lack of abnormalities. Rabbits were housed in stainless steel and polycarbonate cages (Techniplastic, West Chester, PA, USA), at 18–21 °C, with 30–70% humidity, and a 12-h: 12-h light: dark cycle (lights on at 0600). Rabbits were fed 250 g of a commercial pelleted rabbit diet (diet 2030, Harlan Laboratories, Madison, WI, USA) twice daily and were allowed free access to municipal water via an automatic watering system (Edstrom Industries, Waterford, WI, USA). After one-week of acclimatization, rabbits were divided into control (4 animals) and treated (6 animals) groups. The latter animals were injected subcutaneously with 500 µL of purified protein (2 mg/mL) following the polyclonal antibody production protocol of Fishback et al. [31] with some modifications (Figure 6). The study was conducted after obtaining approval from the International Animal Care and Use Committees (IACUCs) IACUC # 30-1Y-0521 (date of approval 10 January 2018).

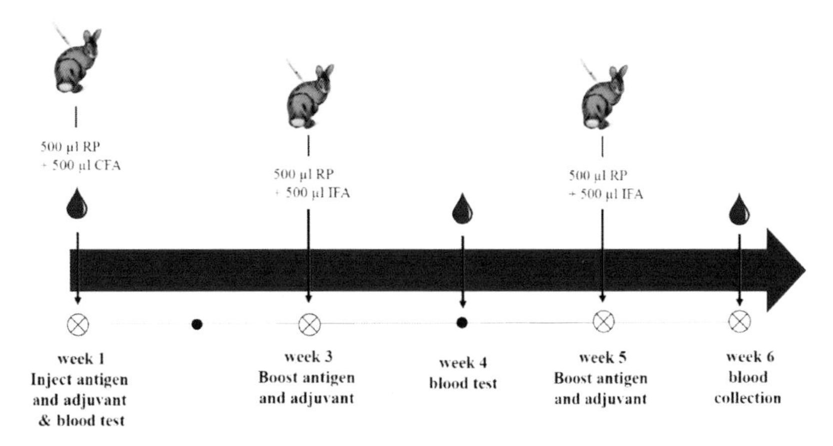

Figure 6. Polyclonal antibody production protocol.

Three milliliter of blood was collected from the auricular artery of each rabbit on weeks 1 and 4 to monitor antibody production. On week 6, under deep anesthesia with a mixture of 22–50 mg/kg ketamine and 5–10 mg/kg xylazine, 3 mL of blood was collected by cardiac puncture. Blood was collected into BD Vacutainer serum separation tubes (BD, Franklin Lakes, NJ, USA), and kept upright at room temperature (20 ± 2 °C) for serum separation following the manufacturer's instructions. Separated sera were stored at −80 °C until further analysis.

Serum IgG Purification and Fractionation

Rabbit sera were obtained by centrifugation of immunized rabbit blood at 4000 rpm for 5 min at 4 °C. IgG fractions were obtained by loading sera onto an affinity Protein G-Sepharose column. In brief, the IgG1 fraction was eluted with glycine buffer, pH 2.7, and the IgG3 fraction obtained by elution with glycine buffer, pH 3.5. All IgG fractions were immediately neutralized in a neutralization buffer (1 M Tris–HCl, pH 8.0, 150 mM NaCl, 5 mM EDTA) [32].

Labelling of Antibodies

Ten mg of alkaline phosphatase (AP) were mixed with purified IgGs (2.5 mg) in 5 mL of 50 mM phosphate buffer, pH 6.8. Mixtures were dialyzed against 2 L of 50 mM phosphate buffer for 24 h at 4 °C. One mL of 1.25% glutaraldehyde was added to each mixture and gently stirred for 2 h at room temperature (20 °C ± 2). Two hundred fifty µL of 0.2 M glycine solution was added followed by further stirring for 2 h. Mixtures then were dialyzed twice against 2 L of 1.0× PBS containing 1 mM magnesium chloride, followed by centrifugation at 10,000 rpm for 5 min to remove any precipitate [33]. Each conjugate was further purified on a Sephacryl S200 column (5 × 150 mm, GE Health care, Danderyd, Sweden) previously equilibrated with PBS and eluted with the same buffer.

4.5.5. Quality Checks

An indirect enzyme-linked immunosorbent assay (I-ELISA) was used to detect aflatoxin B1 in food samples using polyclonal antibodies. Antibodies were compared using an antiserum produced by Sigma (Berlin, Germany). One gram of contaminated food sample was extracted in 10 mL coating buffer. One hundred microliter of sample extract was added to each well. Plates were then incubated at 37 °C for 3 h and blocked with 200 µL of blocking buffer (1× PBS and 0.5% BSA) for 1 h at room temperature (20 °C ± 2). One hundred microlite concentration of 1:800 diluted secondary antibody alkaline phosphatase-conjugated (anti-rabbit antibody) was added and the mixture was incubated at 37 °C for 1 h. All washing steps between incubations used 1× PBS-T buffer. Finally, freshly prepared pNPP substrate was added; plates were incubated at room temperature for 30 min away from direct light, and the absorbance was measured at 405 nm. All experimental steps are summarized in Figure 7.

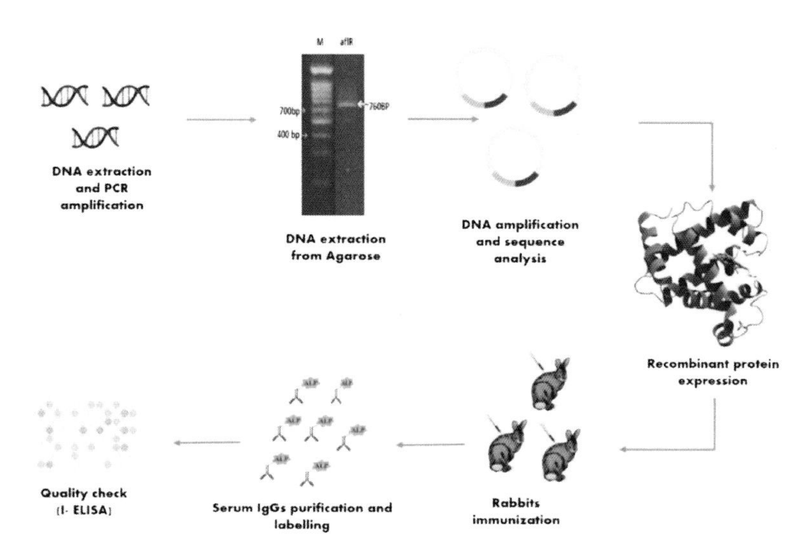

Figure 7. Summary of experimental steps.

5. Statistical Analysis

Data were statistically analyzed using SPSS software (version 16). One-way analysis of variance was used to assess the significance of differences among means, with a significance threshold of $p < 0.05$. The Pearson correlation coefficient (r) was also calculated ($p < 0.001$) to assess the strength of linear relationships between variables.

Author Contributions: Conceptualization, E.H., N.M.A.E.-A., A.M.G.D., M.G.S., A.A.I., A.M.E. and A.N.B.; methodology, E.H., N.M.A.E.-A., A.M.G.D., M.G.S., A.A.I. and A.M.E.; software and static analysis, E.H., N.M.A.E.-A., A.M.G.D., M.G.S., A.A.I., A.M.E. and A.N.B.; formal analysis, E.H., N.M.A.E.-A., A.M.G.D., M.G.S. and A.N.B.; investigation and data curation, E.H., N.M.A.E.-A., A.M.G.D., M.G.S. and A.N.B.; resources, E.H., N.M.A.E.-A., A.M.G.D. and M.G.S.; writing—original draft preparation, E.H., N.M.A.E.-A., A.M.G.D. and M.G.S.; writing—review and editing, E.H., N.M.A.E.-A., A.M.G.D., M.G.S., A.A.I., A.M.E. and A.N.B. All authors have read and agreed to the published version of the manuscript.

Funding: This research received no external funding.

Institutional Review Board Statement: Animal experiment was conducted upon the approval of International Animal Care and Use Committees (IACUCs) (Permission number: IACUC # 30-1Y-0521, date of approval 10 January 2018).

Informed Consent Statement: Not applicable.

Conflicts of Interest: The authors declare no conflict of interest.

References

1. Goud, K.Y.; Reddy, K.K.; Satyanarayana, M.; Kummari, S.; Gobi, K.V. A review on recent developments in optical and electrochemical aptamer-based assays for mycotoxins using advanced nanomaterials. *Microchim. Acta* **2019**, *187*, 29. [CrossRef]
2. Hussain, K.K.; Malavia, D.; Johnson, E.M.; Littlechild, J.; Winlove, C.P.; Vollmer, F.; Gow, N.A.R. Biosensors and Diagnostics for Fungal Detection. *J. Fungi* **2020**, *6*, 349. [CrossRef] [PubMed]
3. Bryła, M.; Waśkiewicz, A.; Podolska, G.; Szymczyk, K.; Jędrzejczak, R.; Damaziak, K.; Sułek, A. Occurrence of 26 Mycotoxins in the Grain of Cereals Cultivated in Poland. *Toxins* **2016**, *8*, 160. [CrossRef] [PubMed]
4. Ezekiel, C.; Sulyok, M.; Babalola, D.; Warth, B.; Ezekiel, V.; Krska, R. Incidence and consumer awareness of toxigenic Aspergillus section Flavi and aflatoxin B1 in peanut cake from Nigeria. *Food Control* **2013**, *30*, 596–601. [CrossRef]
5. Dai, Y.; Huang, K.; Zhang, B.; Zhu, L.; Xu, W. Aflatoxin B1-induced epigenetic alterations: An overview. *Food Chem. Toxicol.* **2017**, *109*, 683–689. [CrossRef] [PubMed]
6. Payne, G.A.; Widstrom, N.W. Aflatoxin in maize. *Crit. Rev. Plant Sci.* **1992**, *10*, 423–440. [CrossRef]

7. Shotwell, O.L.; Goulden, M.L.; Hesseltine, C.W. Aflatoxin: Distribution incontaminated corn. *Cereal Chem.* **1974**, *51*, 492–499.
8. Filazi, A.; Tansel, U. Occurrence of Aflatoxins in Food. *Aflatoxins Recent Adv. Future Prospect.* **2013**, 143–170. [CrossRef]
9. Herzallah, S.M. Determination of aflatoxins in eggs, milk, meat and meat products using HPLC fluorescent and UV detectors. *Food Chem.* **2009**, *114*, 1141–1146. [CrossRef]
10. Wagner, K.; Springer, B.; Pires, V.P.; Keller, P.M. Molecular detection of fungal pathogens in clinical specimens by 18S rDNA high-throughput screening in comparison to ITS PCR and culture. *Sci. Rep.* **2018**, *8*, 1–7. [CrossRef]
11. Gallo, A.; Stea, G.; Battilani, P.; Logrieco, A.F.; Perrone, G. Molecular characterization of Aspergillus flavus population isolated from maize during the first outbreak of aflatoxin contamination in Italy. *Phytopathol. Mediterr.* **2012**, *51*, 198–206. [CrossRef]
12. Alahi, M.E.E.; Mukhopadhyay, S.C. Detection methodologies for pathogen and toxins: A review. *Sensors* **2017**, *17*, 1885. [CrossRef]
13. Maggira, M.; Ioannidou, M.; Sakaridis, I.; Samouris, G. Determination of Aflatoxin M1 in Raw Milk Using an HPLC-FL Method in Comparison with Commercial ELISA Kits—Application in Raw Milk Samples from Various Regions of Greece. *Vet. Sci.* **2021**, *8*, 46. [CrossRef]
14. Vosough, M.; Bayat, M.; Salemi, A. Matrix-free analysis of aflatoxins in pistachio nuts using parallel factor modeling of liquid chromatography diode-array detection data. *Anal. Chim. Acta* **2010**, *663*, 11–18. [CrossRef] [PubMed]
15. Sapsford, K.; Ngundi, M.; Moore, M.; Lassman, M.; Shriver-Lake, L.; Taitt, C.; Ligler, F. Rapid detection of food-borne contaminants using an array biosensor. *Sens. Actuators B Chem.* **2006**, *113*, 599–607. [CrossRef]
16. Kolaskar, A.S.; Tongaonkar, P.C. A semi-empirical method for prediction of antigenic determinants on protein antigens. *FEBS Lett.* **1990**, *276*, 172–174. [CrossRef]
17. Schmidt, A.M. Development and application of synthetic peptides as vaccines. *Biotechnol. Adv.* **1989**, *7*, 187–213. [CrossRef]
18. Mahan, D.E.; Morrison, L.; Watson, L.; Haugneland, L.S. Phase change enzyme immunoassay. *Anal. Biochem.* **1987**, *162*, 163–170. [CrossRef]
19. Gruzdys, V.; Cahoon, K.; Pearson, L.; Lehman, C.M. Method Verification Shows a Negative Bias between 2 Procalcitonin Methods at Medical Decision Concentrations. *J. Appl. Lab. Med.* **2019**, *4*, 69–77. [CrossRef]
20. Harrison, B.D.; Barker, H.; Bock, K.R.; Guthrie, E.J.; Meredith, G.; Atkinson, M. Plant viruses with circular single-stranded DNA. *Nat. Cell Biol.* **1977**, *270*, 760–762. [CrossRef]
21. Padidam, M.; Beachy, R.N.; Fauquet, C.M. Classification and identific-ation of geminiviruses using sequence comparison. *J. Gen. Virol.* **1995**, *76*, 249–263. [CrossRef]
22. Abdel-Salam, A.M. Isolation and characterization of a whitefly-transmitted geminivirus associated with the leaf curl and mosaic symptoms on cotton in Egypt. *Arab. J. Biotech.* **1999**, *2*, 193–218.
23. Whitaker, T.B. Detecting Mycotoxins in Agricultural Commodities. *Mol. Biotechnol.* **2003**, *23*, 61–72. [CrossRef]
24. Lee, N.A.; Rachaputi, N.C.; Wright, G.C.; Krosch, S.; Norman, K.; Anderson, J.; Ambarwati, S.; Retnowati, I.; Dharmaputra, O.S.; Kennedy, I.R. Validation of analytical parameters of a competitive direct ELISA for aflatoxin B1in peanuts. *Food Agric. Immunol.* **2005**, *16*, 149–163. [CrossRef]
25. Lee, S.D.; Yu, I.S.; Jung, K.; Kim, Y.S. Incidence and Level of Aflatoxins Contamination in Medicinal Plants in Korea. *Mycobiology* **2014**, *42*, 339–345. [CrossRef]
26. Scaglioni, P.T.; Badiale-Furlong, E. Rice husk as an adsorbent: A new analytical approach to determine aflatoxins in milk. *Talanta* **2016**, *152*, 423–431. [CrossRef]
27. Ertekin, Ö.; Pirinçci, Ş.Ş.; Öztürk, S. Monoclonal IgA Antibodies for Aflatoxin Immunoassays. *Toxins* **2016**, *8*, 148. [CrossRef]
28. Saiki, R.K.; Gelfand, D.H.; Stoffel, S.; Scharf, S.J.; Higuchi, R.; Horn, G.T.; Mullis, K.B.; Erlich, H.A. Primer-directed enzymatic amplification of DNA with a thermostable DNA polymerase. *Science* **1988**, *239*, 487–491. [CrossRef] [PubMed]
29. Rosano, G.L.; Ceccarelli, E.A. Recombinant protein expression in Escherichia coli: Advances and challenges. *Front. Microbiol.* **2014**, *5*, 172. [CrossRef] [PubMed]
30. Laemmli, U.K. Cleavage of Structural Proteins during the Assembly of the Head of Bacteriophage T4. *Nature* **1970**, *227*, 680–685. [CrossRef] [PubMed]
31. Fishback, J.E.; Stronsky, S.M.; Green, C.A.; Bean, K.D.; Froude, J.W. Erratum: Antibody production in rabbits administered Freund's complete adjuvant and carprofen concurrently. *Lab Anim.* **2016**, *45*, 121. [CrossRef] [PubMed]
32. Gianazza, E.; Arnaud, P. A general method for fractionation of plasma proteins. Dye-ligand affinity chromatography on immobilized Cibacron blue F3-GA. *Biochem. J.* **1982**, *201*, 129–136. [CrossRef] [PubMed]
33. Avrameas, S. Coupling of enzymes to proteins with glutaraldehyde: Use of the conjugates for the detection of antigens and antibodies. *Immunochemistry* **1969**, *6*, 43–52. [CrossRef]

 MDPI

Article

Undertaking a New Regulatory Challenge: Monitoring of Ergot Alkaloids in Italian Food Commodities

Veronica Maria Teresa Lattanzio [1], Emanuela Verdini [2], Stefano Sdogati [2], Angela Caporali [2], Biancamaria Ciasca [1] and Ivan Pecorelli [2,*]

[1] National Research Council of Italy, Institute of Sciences of Food Production, Via Amendola 122/O, 70126 Bari, Italy; veronica.lattanzio@ispa.cnr.it (V.M.T.L.); biancamaria.ciasca@ispa.cnr.it (B.C.)
[2] Chemistry Department, Istituto Zooprofilattico Sperimentale dell'Umbria e delle Marche "Togo Rosati", 06126 Perugia, Italy; e.verdini@izsum.it (E.V.); stefano.sdogati@izsum.it (S.S.); a.caporali@izsum.it (A.C.)
* Correspondence: i.pecorelli@izsum.it

Abstract: The present manuscript reports on monitoring data of 12 ergot alkaloids (EAs) in cereal and cereal-derived products, collected in Italy over the period 2017–2020, for official control purposes under the edge of the Commission Recommendation 2012/154/EU on the monitoring of the presence of EAs in feed and food. To these purposes, an LC-MS/MS method was set up and applied, after in-house verification of its analytical performance. Besides satisfactory recoveries and precision, the method's quantification limits proved suitable to assess the compliance of cereals and cereal-based foods with the recently issued EU maximum permitted levels (Commission Regulation 2021/1399/EU). The validity of the generated data was also evaluated through the adoption of four proficiency tests, from which acceptable z-score values ($-2 \leq z \leq 2$) were obtained. The method was then applied to analyse a total of 67 samples, collected in Italy over the period 2017–2020. The samples consisted of 18 cereal grains, 16 flours (14 of wheat and 2 of spelt) and 31 other types of cereals derivatives (including 9 for infants). Overall, the EAs analysis returned a high percentage of left-censored data (>86%). Among the positive samples, the highest contamination levels, up to 94.2 μg/kg, were found for ergocristine (12% incidence), followed by ergocristinine (7% incidence) with levels of up to 48.3 μg/kg.

Keywords: ergot alkaloids; LC-MS/MS method; wheat; cereal products; occurrence

Key Contribution: A new regulation, setting the EU's maximum permitted level for 12 EAs has been issued (Commission Regulation 2021/1399/EU). Data were collected for 12 EAs in cereals and derived products. A fit-for-purpose LC-MS/MS method was validated. Among the positive samples, the highest contamination levels—up to 94.2 μg/kg—were found for ergocristine.

Citation: Lattanzio, V.M.T.; Verdini, E.; Sdogati, S.; Caporali, A.; Ciasca, B.; Pecorelli, I. Undertaking a New Regulatory Challenge: Monitoring of Ergot Alkaloids in Italian Food Commodities. *Toxins* **2021**, *13*, 871. https://doi.org/10.3390/toxins13120871

Received: 28 October 2021
Accepted: 29 November 2021
Published: 6 December 2021

Publisher's Note: MDPI stays neutral with regard to jurisdictional claims in published maps and institutional affiliations.

1. Introduction

The EAs are mycotoxins produced by several species of fungi in the genus *Claviceps*. In Europe *Claviceps purpurea* is the most widespread and it commonly affects cereals such as rye, wheat, triticale, barley, millets and oats [1]. During fungi infection, healthy kernels are replaced by dark mycelial masses known as sclerotia (also known as ergots, or ergot bodies) that contain high concentrations of various EAs [2].

The toxicity of EAs is well known and has been characterized [3,4]. Though some are cytotoxic and antimicrobial, most are primarily neurotropic. Today, ergotism has practically been eliminated as a human disease, but it remains an important veterinary problem, particularly in cattle, horses, sheep, pigs and poultry [5].

Based on the twelve EAs predominantly present in the sclerotia of *C. purpurea*, the EFSA Panel on Contaminants in the Food Chain (CONTAM Panel) concluded that chemical analysis should focus mainly on ergometrine (EM), ergometrinine (EMI), ergosine (ES), ergosinine (ESI), ergotamine (ET), ergotaminine (ETI), ergocornine (EC), ergocorninine

(ECI), mixture of α- and β-isomers of ergocryptine (EKR) and ergocryptinine (EKRI), ergocristine (ECR) and ergocristinine (ECRI) (Figure 1). The -inine epimers are described to be biologically inactive, however, an interconversion occurs under alkaline or acidic conditions and, thus, the CONTAM Panel based its risk assessment on both forms (-ine and -inine) [3].

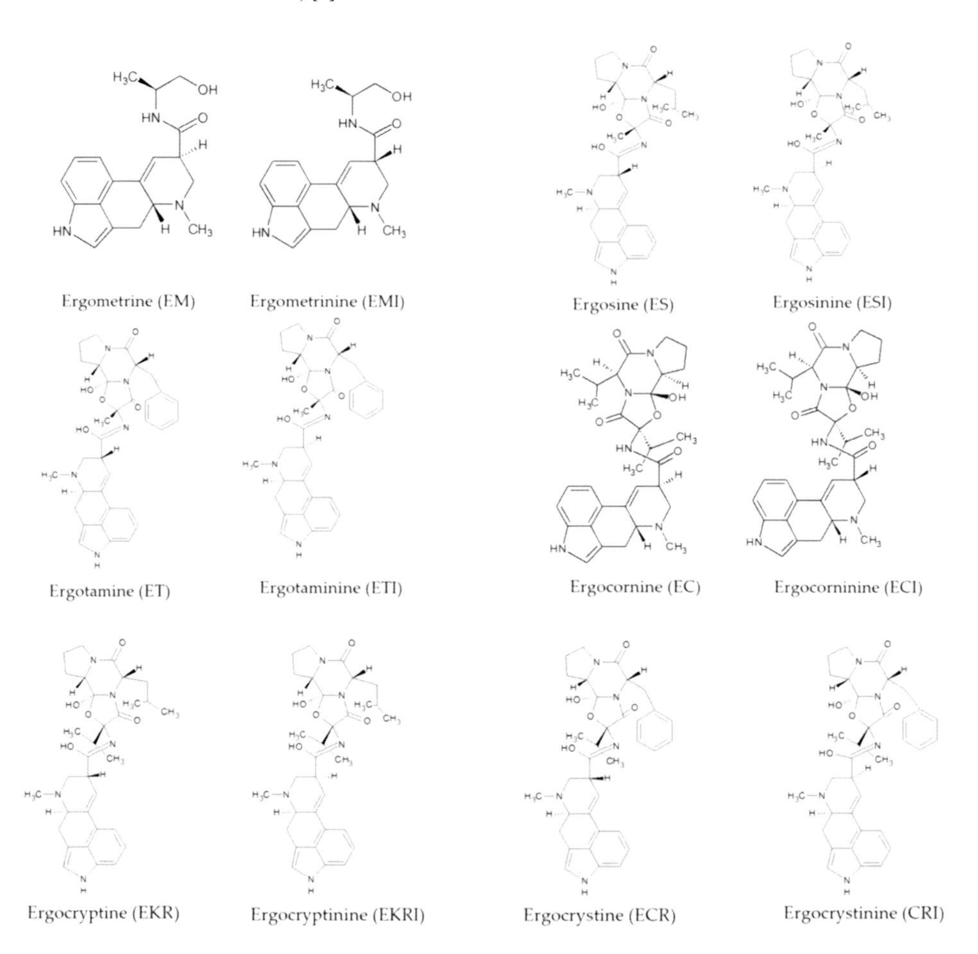

Figure 1. Structure of ergot alkaloids.

Although, today, advanced cleaning procedures prior to milling are rather effective, EAs are still found in food and feed commodities, sometimes at relatively high levels [6,7]. The occurrence data on EAs in food and feed submitted to EFSA indicates that ET, ECR, ES and EC mostly contribute to the overall content of EAs. Furthermore, the highest concentrations of EAs were reported for rye (grains, milling) products and by-products [3,4]. The CONTAM Panel recommended, that efforts should continue to collect more data on occurrence of the above EAs in relevant food and feed commodities. Special attention should be paid to processed food and to specific foods consumed by vegetarians or raw grain consumers. Moreover, the CONTAM Panel underlined the need for commercially available reference standards, such as for isotope-labelled internal standards and certified reference materials (CRM) for the analysis of EAs.

As a follow up to the conclusions and the information contained in the EFSA opinion, a Commission Recommendation on the monitoring of the presence of EAs in feed and food has been in force since 2012 to stimulate analytical data collection regarding the occurrence of EAs identified in the EFSA opinion in relevant food and feed commodities. Furthermore, the Commission Recommendation encouraged the collection of specific information on the relationship between the presence of ergot sclerotia and the level of individual EAs in food and feed in order to set appropriate limits [8]. Finally, in response to EFSA Recommendation, regarding "harmonised performance criteria for the analysis of EAs in feed and food" [3], the Committee agreed that the method of analysis used for the monitoring of ergot alkaloids should have a limit of quantification (LOQ) of 20 µg/kg as a minimum acceptable criterion, but preferably, this value should be 10 µg/kg or lower [9].

Recently, on 24 August 2021, the European Commission published Regulation (EU) 2021/1399 [10], amending Regulation (EC) 1881/2006 [11]. The new Regulation sets the maximum permitted limits for the sum of the above mentioned 12 EAs as lower-bound concentrations (i.e., calculated on the assumption that all values of the different individual ergot alkaloids below the limit of quantification are equal to zero) in cereal-based food products (Table 1). The limits for these alkaloids relate to barley, wheat, spelt, oats, rye and processed cereal-based foods for infants and children and will apply from 1 January 2022. In the Regulation, a higher maximum permitted level is set for milling products containing bran (identified on the base of ash content) taking into account the absorption by cereals of dust containing high levels of EAs.

Table 1. Maximum permitted level for ergot alkaloids in food by the European Commission (Commission Regulation (EU) 2021/1399). [a] Effective from 1 July 2024.

Foodstuff	Maximum Level for the Sum of 12 EAs µg/kg)
milling products of barley, wheat, spelt, oats grains(with an ash content lower than 900 mg/100 g)	100 (50 [a])
milling products of barley, wheat, spelt, oats (with an ash content equal or higher than 900 mg/100 g)	150
barley, wheat, spelt and oats grains placed on the market for the final consumer	150
rye milling products and rye placed on the market for the final consumer	500 (250 [a])
wheat gluten	400
processed cereal based food for infants and young children	20

The maximum levels of EAs set in Regulation (EU) 2021/1399 imply that the analytical methods, for enforcement purposes, should have a LOQ lower than the value previously established by the EU document [9]. Specifically, if calculated according to the formula reported in the UNI CEN/TR 16059:2010 [12] the LOQ for monitoring of milling products, bran milling products/grain for human consumption for cereal other than rye and processed cereal-based baby foods shall be equal to 4.0, 6.0 and 0.8 µg/kg per each ergot compound, respectively.

Different methods have been reported in the literature for the analysis of ergot alkaloids, mainly liquid chromatographic methods coupled to fluorometric or tandem mass-spectrometric detectors (FLD or MS/MS). A critical review can be found in Chung 2021 [13], discussing the advantages and disadvantages of available methods for determination of EAs in cereals and feed, covering the period from 2008 to 2020. The review points out that, although both LC-FLD and LC-MS/MS can be used for the analysis of the 12 EU-recommended EAs, the latter has a greater sensitivity, but it is affected by a matrix effect especially, for EM and EMI. Another analytical challenge, stressed in the review, is the co-elution of alpha and beta isomers of EKR (α-EKR, β-EKR) for most of the reported methods, due to the use of C18 analytical columns. A proficiency test, conducted in 2017, revealed that an acceptable resolution was obtained with phenyl-hexyl as a stationary phase [14]. Finally, the review underlines that only very few reported methods can fulfil the regulated LOQs for individual epimers in processed cereal-based food for infants and young children owing to its lower limit. Recently a modified QuEChERS-based method coupled to LC-MS/MS as a detection technique was successfully validated for the detection and quantification of EAs in dry cereal-based baby foods with individual LOQs of 0.5 μg/kg [15], however the method did not provide the separation of the α and β isomers of EKR and EKRI.

A standardized method, for the determination of EAs in cereals and cereal products by dispersive solid phase extraction (dSPE) clean-up and LC-MS/MS, has been recently issued by the European Committee for Standardization for official control purposes [16]. The method has been validated in the range of 13.2 μg/kg to 168 μg/kg for the sum of the twelve EAs, in rye flour, rye bread and cereal products (breakfast cereals, infant breakfast cereals and crispbread) that contained rye as an ingredient, as well as seeded wholemeal flour and a barley and rye flour mixture. Method performances were satisfactory in the range 24.1 μg/kg to 168 μg/kg for sum of EAs, whereas for concentrations below 24.1 μg/kg the method resulted to be only suitable for screening purposes.

Due to analytical challenges in the EAs determination, the occurrence of data available in the literature are scarce and provide a limited picture of EAs distribution worldwide.

The present manuscript reports on monitoring data of EAs in cereal and cereal-derived products collected in Italy over the period 2017–2020, as requested by the national implementation of the monitoring recommendations [8]. To these purposes a LC-MS/MS method for the determination of EAs in cereal and derived products has been optimized and in house validated to verify its fitness for purpose. Validation data will be reported and discussed, also taking into account the recently issued Regulation requirements.

2. Results and Discussion

2.1. Method Set Up and In-House Validation

The aim of this work was to set up and validate a fit-for-purpose method for the routine monitoring of EA in official control. Since the CEN standard [16] was not yet available at the time of the study, a new method was set up, starting from the procedure developed by Kokkonen et al. 2010 (https://doi.org/10.1002/jssc.201000114, accessed on November 2021). The primary improvements adopted to make the method suitable for routine analysis were a shorter extraction time (shaker time of 30 min vs. 60 min) and the use of a calibration curve, in the mobile phase, instead of a matrix-assisted calibration curve for quantification. This last point was very important for official controls, considering that different types of food products are generally analysed in the same batch.

Prior to the validation study, the chromatographic separation of target EAs, was optimized. Special attention was paid to EKR and EKRI, which have been shown to be particularly challenging under conventional reverse-phase chromatographic conditions, leading to chromatographically unresolved double peaks for both compounds, corresponding to the α- and β-forms [14].

Within this study, two different reverse-phase columns were selected and tested to improve EKR and EKRI separation: a Kinetex EVO C18 (100 × 2.1 mm, i.d. 2.6 μm) and an Acquity UPLC BEH C-18 (150 × 2.1 mm, i.d. 1.7 μm). Complete separation of 12 EAs was achieved using column Acquity UPLC BEH C-18 as reported in Figure 2.

Both columns were able to separate α and β isomers of EKR, while for EKRI, the Kinetex column did not provide any separation. For this reason, the Acquity UPLC BEH C-18 column was chosen. The EKR and EKRI results are shown in Figure 3.

Although the separation of the a- and β-isomers, co-occurring in real samples, would be desirable, a joint quantification (estimating the sum expressed as α isomer) might still be acceptable, in routine monitoring, considering the lack of available reference standards for the β forms.

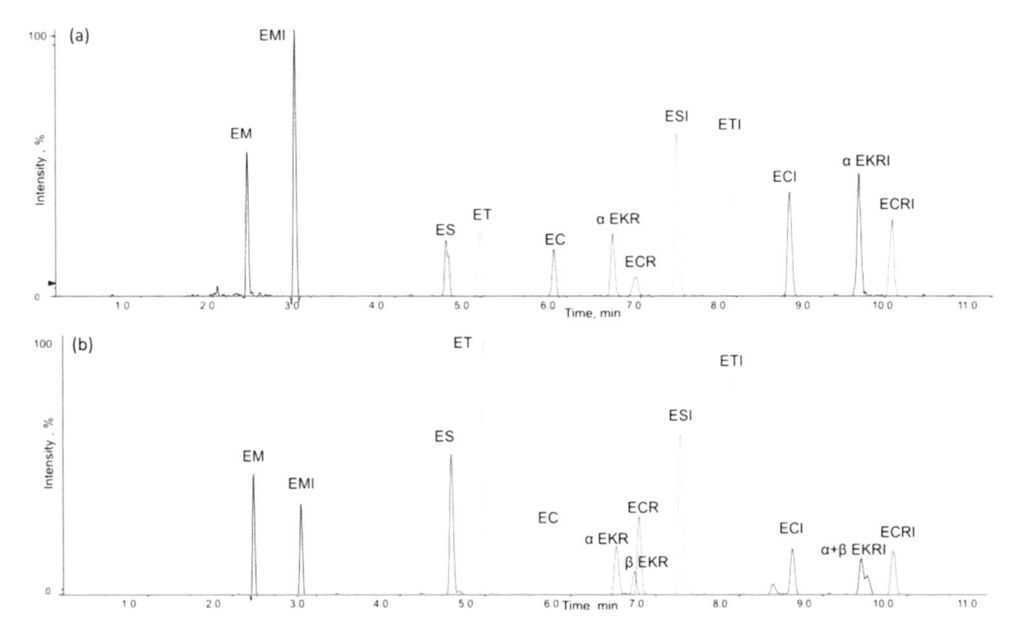

Figure 2. LC-MS/MS chromatogram of (**a**) wheat sample spiked with 2.5 μg/kg of each EA (EKR and EKRI alpha isomer only) and (**b**) naturally contaminated barley sample with EM (127 μg/kg), EMI (50 μg/kg), ESI (197 μg/kg), ET (858 μg/kg), ETI (209 μg/kg), EC (266 μg/kg), ECI (141 μg/kg), sum of α + β EKR (262 μg/kg), sum of α + β EKRI (119 μg/kg), ECR (459 μg/kg) and ECRI (161 μg/kg).

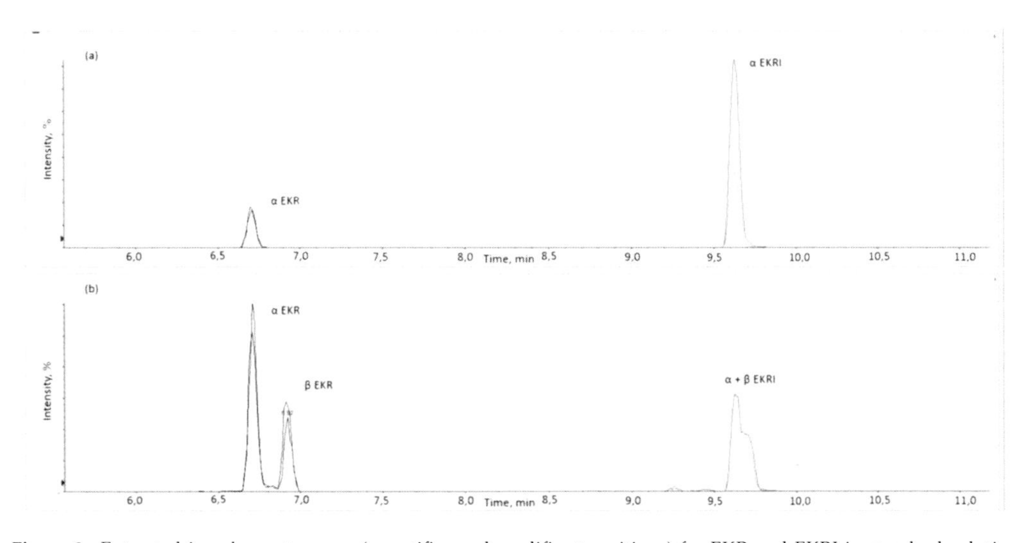

Figure 3. Extracted ion chromatograms (quantifier and qualifier transitions) for EKR and EKRI in standard solution (0.04 μg/mL) (**a**) and wheat sample naturally contaminated with EKR (mixture of α-EKR and β-EKR) (351 μg/kg) and EKRI (141 μg/kg) analysed for proficiency test using an Aquity BEH C-18 column (**b**).

Given that EAs are more likely occurring in cereals and relevant derived products, the in-house method's performance was evaluated with wheat at concentrations as low as possible (e.g., the estimated LOQ) and at higher levels, taking into account available occurrence data.

Data obtained from in-house validation with wheat are summarized in Table 2.

Table 2. In-house analytical performances of the LC-MS/MS method for EAs, including spiking levels, limits of detection (LOD) and quantitation (LOQ), average recovery %, repeatability (RSDr) and within-laboratory reproducibility (RSD_{WLR}). [a] Spiking levels were set at LOQ (2.5 µg/kg), 2xLOQ (5 µg/kg) and 4xLOQ (10 µg/kg).

	LOD (µg/kg)	LOQ (µg/kg)	Spiking Level [a] (µg/kg)	Mean Recovery, (%)	RSDr, (%)	RSD_{WLR}, (%)
EM	0.3	0.8	2.5 5 10	97 99 108	6 8 8	7 8 8
EMI	0.2	0.6	2.5 5 10	111 112 119	7 8 5	11 8 5
ES	0.3	0.9	2.5 5 10	103 101 114	8 9 8	13 13 8
ESI	0.3	0.9	2.5 5 10	103 110 105	7 7 9	7 8 9
ET	0.3	1.1	2.5 5 10	105 100 105	8 5 10	8 5 11
ETI	0.2	0.7	2.5 5 10	111 109 113	8 6 4	10 6 4
EC	0.4	1.2	2.5 5 10	105 95 105	9 13 9	9 13 9
ECI	0.2	0.7	2.5 5 10	97 97 106	8 11 7	8 11 7
α EKR	0.7	2.1	2.5 5 10	105 95 104	8 11 7	10 15 9
α EKRI	0.2	0.8	2.5 5 10	87 96 100	12 7 8	12 12 8
ECR	0.7	2.3	2.5 5 10	105 93 108	10 13 13	10 14 14
ECRI	0.4	1.2	2.5 5 10	94 99 107	8 8 5	8 8 5

Abbreviation: ergometrine (EM), ergometrinine (EMI), ergosine (ES), ergosinine (ESI), ergotamine (ET), ergotaminine (ETI), ergocornine (EC), ergocorninine (ECI), α isomers of ergocryptine (α EKR), α isomers of ergocryptinine (α EKRI), ergocristine (ECR) and ergocristinine (ECRI).

Taking into account that no acceptability criteria for linearity were set in the EU legislation regarding performance criteria for mycotoxins analysis, the authors used residuals to evaluate linearity, and the criteria was met for all 12 compounds [17].

The estimated LOQs (see Section 4.6.1) ranged from 0.6 to 2.3 µg/kg for each compound and were compliant with CEN TR 16059 criteria. According to this guideline, when the legal maximum limit (ML) is set for a sum, the LOQs suitable for enforcement of the legal limit shall be equal to or less than ML divided by 2n (where n is the number of compounds involved). Therefore, the desired values for the monitoring of wheat-milling products and bran-milling products/grain for human consumption (other than rye) shall be set at 4.0 and 6.0 µg/kg for each ergot compound respectively. The values calculated according to the Guide are above the LOQs obtained for the present method.

Then, taking into account the experimentally determined LOQs values, the lowest validation level was set at 2.5 µg/kg for each individual toxin, whereas the others were set at 2 and 5 × LOQ. Mean recoveries ranged from 87 to 119%, whereas repeatability (RSD_r) and within-laboratory reproducibility (RSD_{WLR}) were lower than 13% and 15%, respectively (Table 2). Overall, very satisfactory performances were obtained for the proposed method.

A further confirmation of the reliability of the results obtained with the validated method should be sought in the positive outcome of the participation in four Proficiency Tests (FAPAS 22158, Rye Flour 2019, individual EA range 3–65 µg/kg; Bipea 99-1 Barley 2020, individual EA range 50–858 µg/kg; Bipea 99-2 Rye 2020, individual EA range 18–329 µg/kg; Bipea 99-3 Wheat 2020, individual EA range 76–1030 µg/kg).

Acceptable z-score values ($-2 \leq z \leq 2$) were obtained in all PTs (for a total of $n = 49$ provided results), even in cases where the values of the individual molecules were close to or even slightly lower than the estimated method LOQs.

2.2. Applicability of the New Method for Official Control Purposes

To provide evidence of the applicability and fitness for purpose of the presented method for official controls, data generated within the Italian national monitoring program on the period 2017–2020 are reported herein. Occurrence data for EAs are summarized in Table 3, whereas individual data for each toxin in all analysed samples are provided as Supplementary Material (Table S1).

Table 3. Concentration of EAs in cereal grains and cereal products (67 samples analysed). [a] Values calculated on positive samples. LC (left-censored data).

	Incidence	Mean [a] (µg/kg)	Range (µg/kg)	LC
EM	13	10.2	2.5–25	87
EMI	4	4.5	2.5–7.9	96
ES	10	7.4	2.5–23.5	90
ESI	4	4.7	2.5–6.2	96
ET	7	6.7	2.5–6.1	93
ETI	3	6.1	2.5–9.7	97
EC	6	8.8	2.5–13.9	94
ECI	3	7.5	2.5–12.4	97
EKR	7	9.5	2.5–27.8	93
EKRI	4	8.0	2.5–19.0	96
ECR	12	16.3	2.5–94.2	88
ECRI	7	12.4	2.5–48.3	93
Total EAs	25	31.2	2.7–270.7	75

Abbreviations: ergometrine (EM), ergometrinine (EMI), ergosine (ES), ergosinine (ESI), ergotamine (ET), ergotaminine (ETI), ergocornine (EC), ergocorninine (ECI), ergocryptine (EKR), ergocryptinine (EKRI), ergocristine (ECR) and ergocristinine (ECRI).

The analysis of the EAs returned a high percentage of left-censored data (>86%). EM was the most abundant compound, followed by ECR and ES. The individual highest concentration was detected for ECR at 94.2 µg/kg in wheat bran. One sample only (wheat bran) contained all 12 EAs, with a sum of EAs of 271 µg/kg, which could be labelled as non-compliant under the new EU ML of 150 µg/kg [10]. All the other 16 positive samples were compliant, according to the relevant EU ML.

The data in Table 3 were then compared with previously generated ones. The most recent occurrence data for food samples, available in EFSA reports, cover the period 2011 and 2016 and show the highest average contributors to the total concentration to be ET

(18%), ECR (15%) ES (12%) and EM (11%) [4]. Considering the large amount of left-censored data, present in the EFSA data set (86%), to minimize the impact of presence of relatively high LODs/LOQs on the UB (upper bound) scenario, a value of 20 μg/kg was selected as a LOQ cut off for each individual EAs, permitting the exclusion of those samples analysed by methods with poor sensitivity but without excessively compromising the number of available samples. In this respect, the LOQs of the method validated and applied in this manuscript were around 10 times lower than the above cut off level (Table 1) and, for this reason, can be considered fit for the purpose of an accurate occurrence evaluation.

The method was also applied to detect the presence of 12 EAs in cereal products for infants. The method did not report any particular issue; therefore, a future validation in cereal products for infants could be demonstrate its suitability for these product categories.

Available literature data on EAs occurrence in food samples, collected in the period 2015–2021, are summarized in Table 4. Results presented in this work are globally in line with previous studies. EM was also reported as the most common EA in wheat sample from Italy by Debegnac et al. [18], moreover, ECR was predominant in cereal samples from Luxembourg [19] and in French cereals [20].

The literature data provide a limited picture of EAs distribution worldwide. This could be partly attributed to the analytical challenges to be undertaken in analysing EAs. Therefore, the availability of isotopically internal standards could improve the accuracy of quantification. Moreover, the difficult chromatographic separation of alpha and beta EKR and EKRI isomers [14], the carefulness needed in samples and standard management (to avoid the epimerization of EAs during sample treatment) [13,21], make the analysis of EAs very tricky. The highest EA incidence is reported for rye and rye-based products, whereas an incidence lower than 10% was observed for other cereals and derived products, and, therefore, comparable to the data presented herein.

Table 4. Overview of representative studies on the occurrence of EAs in food samples collected worldwide over the period 2015–2021. The selected studies are relevant to data obtained from sets of more than 15 samples.

Country	Food Matrix	N Sample	Incidence %	Mean [a] (μg/kg)	Range (μg/kg)	References
Canada	barley	67	73	1150	2.2–29,425	[22]
Italy	rye-based products	16	7.5	NA	2.6–189	[18]
	wheat-based products	55	47	NA	2.5–1143	
China	cereal samples	123	4	204	9.5–803	[23]
Albania	cereals	228	NA	NA	65–1140	[24]
Algeria	barley	30	4	35.4	18–54	[25]
	wheat	30	8	33.1	3.7–76	
Belgium market	cereal based baby foods	49	49	3.1	0.1–41.6	[15]

N: number of analysed samples, [a] Values calculated on positive samples. NA: Not available in the publication.

Overall, this comparison demonstrated that general applicability of the proposed method and, specifically, that (i) the ranges selected for method's validation encompassed the natural contamination of EAs, not only in Italy, but also in other countries; and (ii) method quantification limits are also suitable to assess EAs contamination in samples for other countries.

The proposed method was suitable to monitor the natural occurrence of EAs in grain and cereal and derived products. Although the method was not validated on cereal products intended for infant consumption, it was applied to the analysis of nine cereal-based food for infants. From the results obtained, the method seems compliant, however further efforts are needed to lower the LOQ.

3. Conclusions

A fit-for-purpose LC-MS/MS method has been developed and validated for the determination of EAs in official control. The method's performances were proven to be suitable in assessing the compliance of cereals and cereal-based foods with the recently issued EU maximum permitted levels (Commission Regulation 2021/1399/EU). Furthermore, the method's applicability was evaluated by implementing it for EAs analysis in the national monitoring program, which included a total of 67 cereal-based samples collected in Italy over the period 2017–2020. Both the generated data and a comparison with previously reported occurrence data indicate that the method's performances, in terms of precision, accuracy, applicability range and quantification limits, are suitable for assessing EAs natural contamination of cereals and derived products.

4. Materials and Methods

4.1. Chemicals and Reagents

EAs were obtained from Romer Labs (Tulln, Austria). Acetic acid and ammonium carbonate were purchased from Honeywell (Wunstorferstrasse, Germany). Ethyl acetate (EtOAc), Methanol (MeOH) and Acetonitrile (ACN) were obtained from Carlo Erba reagent Srl (Milan, Italy). All solvents used were of LC–MS or analytical grade. Water was purified by a Milli-Q system (Millipore, Merck KgaA, Darmstadt, Germany). The MycoSep 150 Ergot columns were purchased from Romer Labs.

4.2. Samples

Sixty-seven official samples were collected in the period between 2017 and 2020 from three six Italian Regions (Umbria, Marche and Puglia) and analyzed by Istituto Zooprofilattico Sperimentale of Umbria and Marche "Togo Rosati". The samples consisted of 18 cereal grains, 16 flours (14 of wheat and 2 of spelt) and 33 other types of cereals derivatives (including 9 for infants) respectively.

Samples were ground by a knife mill (GRINDOMIX GM 300, Restek, Haan, Germany) with dry ice and split in aliquots of 25 g for the analysis. The samples were stored at −20 °C until analysis.

4.3. Reference Materials and Working Solutions

All reference materials (RMs) of EAs were in desiccated form. Reference solutions were prepared by reconstitution, according to the manufacturer's instructions, obtaining a final concentration of 100 μg mL^{-1} for the R epimers of the EAs (ine-epimers) and 25 μg mL^{-1} for S epimers of EAs (inine-epimers) respectively. The obtained RMs solutions were stored in amber vials at −20 °C.

The working solutions (WS) were prepared by dilution of RMs just before use. For EAs-ine epimers an intermediate working mixed solution at 5 μg mL^{-1} was prepared. The intermediate solution of -ine epimers was then combined with single RMs of EAs-inine epimers to obtain a final concentration of 0.5 μg mL^{-1} for each molecule.

4.4. Sample Preparation

Twenty-five grams of sample were weighed in a 250-mL plastic vessel and 100 mL of extraction solution of acetonitrile:ammonium carbonate (200 mg L^{-1} (84:16 v/v)) were added. The samples were mechanically shaken for 30 min. After 15 min of centrifugation at 2780 RCF, 5 mL of extract was collected and loaded into the solid-phase extraction column (MycoSep 150 Ergot). One mL of purified extract was evaporated to dryness at 60 °C under a gentle stream of nitrogen. Finally, the sample was reconstituted with 400 μL of ammonium carbonate solution (200 mg/L)/ACN; (50:50 v/v) and filtered using a 0.2-μm PTFE syringe filter prior to injection into the LC-MS/MS system.

4.5. LC-MS/MS Analysis

The LC-MS/MS instrumental set up consisted of a Nexera X2 UPLC system (LC-30AD binary pump, CTO-20AC column oven and SIL-30AC autosampler, Shimadzu, Kyoto, Japan, 2015) interfaced to an API 3200 Qtrap mass spectrometer (AB Sciex, Foster City, CA, USA, 2009) equipped with an electrospray (ESI) ion source.

The analysis of EAs was performed in positive ionization mode (ESI+), after separation on an Acquity UPLC BEH C-18 (150 × 2.1 mm, i.d. 1.7 μm) connected to a VanGuard (2.1 × 5 mm) both from Waters (Milford, MA, USA). The column oven was set at 40 °C. The flow rate of the mobile phase was 500 μL/min, while the injection volume was 5 μL. Eluent A was a 200-mg/L ammonium carbonate solution and eluent B was acetonitrile. For EAs elution, the starting composition of the eluent was 95% (A) and 5% (B). Then, the following gradient was used: the proportion of eluent B was linearly increased from 5% to 40% over 1 min, then to 50% over the next 3.5 min, then increased to 70%. Finally, it was raised to 99% over 1.5 min and kept constant at 5% for 3 min.

The target mycotoxins were detected in Selected Reaction Monitoring (SRM) mode. The monitored transitions and retention times of single EAs are provided in Table 5. Compliance with SANTE mycotoxin identification criteria for retention time (Rt), chromatographic separation and Ion Ratio (IR) for identification in mass spectrometry was verified (SANTE/12089/2016). Quantification was carried out by external calibration in solvent.

Table 5. Retention times and monitored transitions for individual EAs.

ID	Retention Time (min)	Precursor Ion (*m/z*)	Product Ion (*m/z*)
EM	2.24	326	223 208
EMI	2.82	326	208 223
ES	4.60	548	223 208
ESI	7.30	548	223 208
ET	5.00	582	223 208
ETI	7.95	582	223 277
EC	5.85	562	223 208
ECI	8.65	562	223 277
α EKR	6.55	576	223 268
α EKRI	9.50	576	223 291
ECR	6.80	610	223 268
ECRI	9.85	610	223 208

Abbreviations: ergometrine (EM), ergometrinine (EMI), ergosine (ES), ergosinine (ESI), ergotamine (ET), ergotaminine (ETI), ergocornine (EC), ergocorninine (ECI), α isomers of ergocryptine (α EKR), α isomers of ergocryptinine (α EKRI), ergocristine (ECR) and ergocristinine (ECRI).

4.6. Method Validation Procedure

For the method's validation the following parameters were evaluated: LODs, LOQs, instrumental linearity ranges, recovery rates (%), RSDr and RSD_{WLR}, both using relative standard deviation. All parameters' definitions and acceptability criteria are reported in UNI CEN/TR 16059:2010.

4.6.1. Limit of Detection and Limit of Quantification

LOD value were determined according to the "Estimation of LOD via blank samples" method as reported in the "Guidance Document on the Estimation of LOD and LOQ for Measurements in the Field of Contaminants in Feed and Food" [26]. Specifically, 10 aliquots of a blank matrix were spiked at 1 μg/kg for all EAs. These spiking levels were fixed as low as possible, considering a S/N ratio ≥3 at the expected LOD. The resulting spiked blank samples were analyzed by LC–MS/MS then an LOD and an LOQ were calculated according to Equations (1) and (2) respectively:

$$\mathrm{LOD} = 3.9 * \frac{S_{y,b}}{b} \quad (1)$$

LOD: limit of detection
$S_{y,b}$: standard deviation of the spiked blank signal
b: slope of calibration curve

$$\mathrm{LOQ} = 3.3 * \mathrm{LOD} \quad (2)$$

LOQ: limit of quantification

4.6.2. Linearity Range

Each calibrant solution was prepared by diluting working solutions with acetonitrile/ammonium carbonate solution at 200 mg/L (50:50 *v*/*v*). Calibrant solutions were in the range 0.4–40 ng/mL and were analyzed on three different days over two weeks. Then calibration curve equations were obtained by plotting averaged peak areas vs. concentration of the natural toxin using ordinary least squares (OLS) method, including a (0, 0) point.

The linearity was checked as follows. For each calibration point, y-residuals were obtained by the following Equation (3)

$$y - \text{residuals} = y_i - \hat{y}_i \quad (3)$$

where y_i are experimental values used for the regression equation calculation and $\hat{y}_i$ values are the points on the calculated regression line corresponding to individual x-values.

When, for all points, the residual along y axis were ≤±20%, the calibration curve was considered linear [17].

4.6.3. Recovery, Repeatability and Within-Laboratory Reproducibility

Recoveries, RSDr and RSD_{WLR} for each molecule were evaluated according to UNI CEN/TR 16059:2010.

EAs validation was performed in wheat at three mass fraction levels, specifically 2.5, 5 and 10 μg kg^{-1} (corresponding to LOQ, 2xLOQ and 4xLOQ respectively) on two different days by two independent operators under repeatability conditions (eight replicates each). To obtain the WLR data, the two groups were combined and recovery% and RSD_{WLR} were calculated as reported in Table 2.

Supplementary Materials: The following are available in online at https://www.mdpi.com/article/10.3390/toxins13120871/s1, Table S1: Occurrence data for Ergot Alkaloids: individual data for each toxin.

Author Contributions: Conceptualization, V.M.T.L. and I.P.; methodology, I.P. and V.M.T.L.; validation, E.V., S.S., A.C. and B.C.; formal analysis, I.P. and V.M.T.L.; resources, I.P.; data curation, B.C., E.V., I.P.; writing—original draft preparation, B.C., E.V.; writing—review and editing, V.M.T.L. and I.P.; supervision, I.P.; funding acquisition, I.P.; All authors have read and agreed to the published version of the manuscript.

Funding: This research was funded by Ministero della salute (Italian Ministry of Health) (IZSUM 08/16 RC).

Institutional Review Board Statement: Not applicable.

Informed Consent Statement: Not applicable.

Data Availability Statement: The data presented in this study are available in Supplementary Material.

Conflicts of Interest: The authors declare no conflict of interest.

References

1. Haarmann, T.; Rolke, Y.; Giesbert, S.; Tudzynski, P. Ergot: From witchcraft to biotechnology. *Mol. Plant Pathol.* **2009**, *10*, 563–577. [CrossRef] [PubMed]
2. Miedaner, T.; Geiger, H.H. Biology, genetics, and management of ergot (*Claviceps* spp.) in rye, sorghum, and pearl millet. *Toxins* **2015**, *7*, 659–678. [CrossRef] [PubMed]
3. EFSA (European Food Safety Authority). Scientific Opinion on Ergot alkaloids in food and feed. *EFSA J.* **2012**, *10*, 2798. [CrossRef]
4. EFSA (European Food Safety Authority); Arcella, D.; Gomez Ruiz, J.A.; Innocenti, M.L.; Roldan, R. Scientific report on human and animal dietary exposure to ergot alkaloids. *EFSA J.* **2017**, *15*, e4902. [CrossRef]
5. Bennett, J.W.; Klich, M. Mycotoxins. *Clin. Microbiol. Rev.* **2003**, *16*, 497–516. [CrossRef] [PubMed]
6. Malysheva, S.V.; Larionova, D.A.; Mavungu, J.D.D.; Saeger, S.D. Pattern and distribution of ergot alkaloids in cereals and cereal products from European countries. *World Mycotoxin J.* **2014**, *7*, 217–230. [CrossRef]
7. López, P.; de Rijk, T.; Sprong, R.C.; Mengelers, M.J.B.; Castenmiller, J.J.M.; Alewijn, M. A mycotoxin-dedicated total diet study in the Netherlands in 2013: Part II–occurrence. *World Mycotoxin J.* **2016**, *9*, 89–108. [CrossRef]
8. EC (European Commission). Commission Recommendation N. 2012/154/EU of 15 March 2012 on the monitoring of the presence of ergot alkaloids in feed and food. *Off. J. Eur. Un.* **2012**, *L77*, 20–21.
9. EC (European Commission). Health and Food Safety Directorate General. Summary Report of the Standing Committee on Plants, Animals, Food and Feed-Section Toxicological Safety of the Food chain. 2013. Available online: https://ec.europa.eu/food/system/files/2016-10/cs_contaminants_catalogue_plant_toxins_compilation_agreed_monitoring_en.pdf(europa.eu) (accessed on 24 October 2021).
10. EC (European Commission). *Commission Regulation (EU) 2021/1399 of 24 August 2021 Amending Regulation (EC) No 1881/2006 as Regards Maximum Levels of Ergot Sclerotia and Ergot Alkaloids in Certain Foodstuffs*; European Commission: Brussels, Belgium, 2021.
11. EC (European Commission). *Commission Regulation (EU) No 1881/2006 of 19 December 2006 Setting Maximum Levels for Certain Contaminants in Foodstuffs*; European Commission: Brussels, Belgium, 2006.
12. Technical report CEN/TR 16059:2010. Food analysis–Performance criteria for single laboratory validated methods of analysis for the determination of Mycotoxins. 2010.
13. Chung, S.W.C. A critical review of analytical methods for ergot alkaloids in cereals and feed and in particular suitability of method performance for regulatory monitoring and epimer-specific quantification. *Food Addit. Contam. Part A Chem. Anal. Control Expo. Risk Assess* **2021**, *38*, 997–1012. [CrossRef]
14. Gonçalves, C.O.; Bouten, K.; Mischke, C.; Bratinova, S.; Stroka, J. *Report on the 2017 Proficiency Test of the European Union Reference Laboratory for Mycotoxins: Determination of Ergot Alkaloids in Rye*; EUR 28946 EN; European Commission: Brussels, Belgium, 2017. [CrossRef]
15. Huybrechts, B.; Malysheva, S.V.; Masquelier, J.A. Targeted UHPLC-MS/MS Method Validated for the Quantification of Ergot Alkaloids in Cereal-Based Baby Food from the Belgian Market. *Toxins* **2021**, *13*, 531. [CrossRef] [PubMed]
16. European Standard EN 17425:2020. *Foodstuffs-Determination of ergot alkaloids in cereals and cereal products by dSPE clean-up and HPLC-MS/MS*; European Committee for Standardization: Brussels, Belgium, 2020.
17. European Commission Directorate General for Health and Food Safety. Document N° SANTE/2015/11945. In *Analytical Quality Control and Method Validation Procedures for Pesticide Residues Analysis in Food and Feed*; European Commission Directorate General for Health and Food Safety: Brussels, Belgium, 2020.
18. Debegnach, F.; Patriarca, S.; Brera, C.; Gregori, E.; Sonego, E.; Moracci, G.; De Santis, B. Ergot Alkaloids in Wheat and Rye Derived Products in Italy. *Foods* **2019**, *8*, 150. [CrossRef] [PubMed]
19. Schummer, C.; Brune, L.; Moris, G. Development of a UHPLC-FLD method for the analysis of ergot alkaloids and application to different types of cereals from Luxembourg. *Mycotoxin Res.* **2018**, *34*, 279–287. [CrossRef] [PubMed]
20. Orlando, B.A.; Maumené, C.; Piraux, F. Ergot and ergot alkaloids in French cereals: Occurrence, pattern and agronomic practices for managing the risk. *World Mycotoxin J.* **2017**, *10*, 327–337. [CrossRef]
21. Schummer, C.; Zandonella, I.; van Nieuwenhuyse, A.; Moris, G. Epimerization of ergot alkaloids in feed. *Heliyon* **2020**, *6*, e04336. [CrossRef] [PubMed]
22. Shi, H.; Schwab, W.; Liu, N.; Yu, P. Major ergot alkaloids in naturally contaminated cool-season barley grain grown under a cold climate condition in western Canada, explored with near-infrared (NIR) and fourier transform mid-infrared (ATR-FT/MIR) spectroscopy. *Food Control* **2019**, *102*, 221–230. [CrossRef]

23. Guo, Q.; Shao, B.; Du, Z.; Zhang, J. Simultaneous Determination of 25 Ergot Alkaloids in Cereal Samples by Ultraperformance Liquid Chromatography-Tandem Mass Spectrometry. *J. Agric. Food Chem.* **2016**, *64*, 7033–7039. [CrossRef] [PubMed]
24. Topi, D.; Jakovac-Strajn, B.; Pavšič-Vrtač, K.; Tavčar-Kalcher, G. Occurrence of ergot alkaloids in wheat from Albania. *Food Addit. Contam. Part A Chem. Anal. Control Expo. Risk Assess.* **2017**, *34*, 1333–1343. [CrossRef] [PubMed]
25. Carbonell-Rozas, L.; Mahdjoubi, C.K.; Arroyo-Manzanares, N.; García-Campaña, A.M.; Gámiz-Gracia, L. Occurrence of Ergot Alkaloids in Barley and Wheat from Algeria. *Toxins* **2021**, *13*, 316. [CrossRef] [PubMed]
26. Wenzl, T.; Johannes, H.; Schaechtele, A.; Robouch, P.; Stroka, J. Guidance Document on the Estimation of LOD and LOQ for Measurements in the Field of Contaminants in Feed and Food. JRC Technical Reports. 2016. Available online: https://ec.europa.eu/jrc/en/publication/guidance-document-estimation-lod-and-loq-measurements-field-contaminants-feed-and-food (accessed on 26 October 2021).

MDPI

Article

Assessment of the Optimum Linker Tethering Site of Alternariol Haptens for Antibody Generation and Immunoassay Development

Luis G. Addante-Moya [1], Antonio Abad-Somovilla [1], Antonio Abad-Fuentes [2], Consuelo Agulló [1] and Josep V. Mercader [2,*]

[1] Department of Organic Chemistry, University of Valencia, Doctor Moliner 50, 46100 Burjassot, Valencia, Spain; luis.addante@uv.es (L.G.A.-M.); antonio.abad@uv.es (A.A.-S.); consuelo.agullo@uv.es (C.A.)

[2] Spanish Council for Scientific Research, Institute of Agrochemistry and Food Technology, Agustí Escardino 7, 46980 Paterna, Valencia, Spain; aabad@iata.csic.es

* Correspondence: jvmercader@iata.csic.es

Abstract: Immunochemical methods for mycotoxin analysis require antigens with well-defined structures and antibodies with outstanding binding properties. Immunoreagents for the mycotoxins alternariol and/or alternariol monomethyl ether have typically been obtained with chemically uncharacterized haptens, and antigen conjugates have most likely been prepared with mixtures of functionalized molecules. For the first time, total synthesis was performed, in the present study, to obtain two haptens with opposite linker attachment locations. The functionalized synthetic haptens were purified and deeply characterized by different spectrometric methods, allowing the preparation of bioconjugates with unequivocal structures. Direct and indirect competitive enzyme-linked immunosorbent assays, using homologous and heterologous conjugates, were employed to extensively evaluate the generated immunoreagents. Antibodies with high affinity were raised from conjugates of both haptens, and a structure-activity relationship between the synthetic haptens and the specificity of the generated antibodies could be established. These results pave the way for the development of novel highly sensitive immunoassays selective of one or two of these *Alternaria* mycotoxins.

Keywords: alternariol; antibody; ELISA; hapten design; immunoassay; linker site

Key Contribution: Two pure regioisomeric alternariol haptens were prepared. Alternariol immunoreagents were generated and characterized.

Citation: Addante-Moya, L.G.; Abad-Somovilla, A.; Abad-Fuentes, A.; Agulló, C.; Mercader, J.V. Assessment of the Optimum Linker Tethering Site of Alternariol Haptens for Antibody Generation and Immunoassay Development. *Toxins* **2021**, *13*, 883. https://doi.org/10.3390/toxins13120883

Received: 9 November 2021
Accepted: 6 December 2021
Published: 10 December 2021

1. Introduction

Alternaria sp. fungi, particularly *A. alternata*, are ubiquitous plant pathogens and saprophytes that infect economically relevant crops such as cereals, vegetables, oilseeds, and fruits. Moreover, these microorganisms can contaminate these commodities after harvest even under refrigeration conditions. They are known to produce a wide variety of toxic secondary metabolites [1], and some of them have been identified by the EFSA Panel on Contaminants in the Food Chain (CONTAM) as a potential risk to human and animal health due to their toxicity and occurrence in food and feed. Surprisingly, there are no specific international regulations for any of the *Alternaria* mycotoxins, and the available data on toxicity, occurrence, and dietary exposure are still limited. In 2011, EFSA carried out the first assessment of the risk of these mycotoxins to human and animal health, based on government and published data [2]. More recently, EFSA conducted a survey on the dietary exposure of European consumers to *Alternaria* toxins [3]. This study found that 8% of these mycotoxins are present in food, with infants and other children being the most exposed population group, and fruit and fruit-based products contributing most to dietary exposure. It is therefore expected that the European Commission will soon set maximum levels for the most common *Alternaria* mycotoxins in foodstuffs.

Alternariol (AOH) and alternariol monomethyl ether (AME), two of the most important compounds belonging to the group of *Alternaria* mycotoxins, appear to be responsible for the teratogenic effects observed in animals. They have also been shown to inhibit in vitro the catalytic topoisomerase activity, which may be associated with human colon and oesophageal cancer [4]. Frequently, these two mycotoxins are found together in samples because they share most of the biosynthetic pathway [5]. AOH and AME have been detected in a wide variety of products, including lentils, carrots, tomatoes, berries, apples, pears, beer, wines, juices, and various grains and flours [6]. To evaluate the relative hazard level of these toxins to human health, a threshold of toxicological concern (TTC) for AOH and AME in adults of 2.5 ng/kg body weight per day was established as a reference parameter by the CONTAM Panel [2]. With limited data available, a 2016 German survey concluded that the percentage of TTC reached by the average adult daily exposure was 1400% and 280% for AOH and AME, respectively [7].

A variety of analytical techniques have been developed for monitoring *Alternaria* toxins in food, including liquid and gas chromatographic methods coupled to several detection systems, as well as different types of immunochemical assays [8,9]. Molecular affinity techniques nowadays represent alternative strategies for rapid, economical, and/or on-site monitoring of mycotoxins. The first antibodies and immunoassays for AOH were reported in 2011 [10,11]. Since then, a few immunoassays have been described using either polyclonal or monoclonal antibodies specific to AOH [12,13], and only one study has been published using an antibody specific to AME [14]. Additionally, Wang et al. have reported a generic antibody for both mycotoxins [15]. In all these studies, the immunogens used to generate antibodies against AOH were made by attaching the mycotoxin to the carrier protein, either directly by a Mannich-type reaction or by carbodiimide-mediated chemistry after nonselective carboxymethylation of the hydroxyl groups. Neither of these methods can be used to guide the position of attachment of the AOH molecular scaffold to the carrier proteins, so the antibodies were actually generated from an undefined mixture of functionalized haptens. More recently, Yao et al. published an immunoassay for AOH using a carboxymethyl hapten to generate polyclonal antibodies [16]. Disappointingly, no synthetic details and spectrometric data of the prepared hapten were provided in that work.

AOH and AME are the two most representative mycotoxins with a tricyclic benzochromenone chemical backbone. Moreover, these compounds contain hydroxyl, methyl, and other substituents in their chemical structure. It is well known that the orientation of the molecule, i.e., the way the hapten is displayed to the immune system, strongly influences the specificity of the generated antibodies. Thus, the synthesis of haptens with the optimal linker tethering site is critical, although it is hard to predict and challenging to perform. Surprisingly, deeply characterized haptens for AOH or AME with unambiguous chemical structures have not been published so far. The aim of the present study was to prepare, purify, and characterize two rationally-designed synthetic haptens of these *Alternaria* mycotoxins with a functionalized linker located at precise sites of the molecule. The ability of these novel immunoreagents to elicit a potent immune response, ultimately leading to high-affinity antibodies, was investigated. In addition, these bioconjugates also allowed the study of the relationship between the functionalization position in the hapten and the specificity of the resulting antibodies.

2. Results and Discussion

2.1. Hapten Design and Synthesis

The generation of antibodies to AOH has so far been based on the preparation of the required immunogens from AOH itself, which does not allow fine control over the specific position of the mycotoxin framework where the functionalized linker is introduced. In this study, we have synthesized two regioisomeric haptens of AOH from scratch. One of them, hapten **AL*a***, incorporates a five-atom carboxylated aliphatic spacer arm through the hydroxyl group at C-9, whereas the other one, hapten **AL*b***, incorporates the same linker via the hydroxyl group at C-3 (Figure 1). In contrast to previous strategies, these two haptens allowed the preparation of bioconjugates with well-defined compositions.

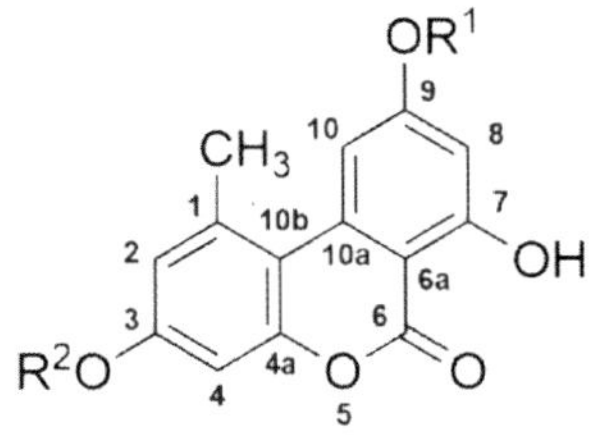

Alternariol: R^1 = H; R^2 = H

Alternariol monomethyl ether: R^1 = CH_3; R^2 = H

Hapten AL*a*: R^1 = $(CH_2)_4CO_2H$; R^2 = H

Hapten AL*b*: R^1 = H; R^2 = $(CH_2)_4CO_2H$

Figure 1. Chemical structures of alternariol, alternariol monomethyl ether, and the synthetic haptens.

The synthetic strategy for preparing hapten **AL*a*** was based on a convergent methodology previously used by several research groups to synthesize AOH and other structurally related molecules [17–20]. A key step in this synthesis is a Pd(0)-catalyzed cross-coupling reaction between an aryl triflate (**4**), which already contained the spacer arm, and an appropriately functionalized arylboronic acid (**6**) (Scheme 1). The aryl triflate **4** was prepared in two steps from the readily available 1,3-benzodioxinone **1** [21,22]. First, an *O*-alkylation reaction with methyl 5-bromovalerate was performed under standard Williamson ether synthesis conditions. The alkylation process produced a 6:1 mixture of di- and mono-*O*-alkylation products, **2** and **3**, respectively, which were easily separated by column chromatography to provide the product resulting from the selective *O*-alkylation of the less hindered hydroxyl group, i.e., **3**, with a 75% yield. The free hydroxyl group of **3** was then converted to the required triflate group by reaction with triflic anhydride in pyridine, giving the triflate **4** in 91% yield. The additional required coupling reactant, aryl boronic acid **6**, was prepared from the orcinol-derived bromide **5** [23] by halogen-metal exchange using butyllithium and reaction of the resulting lithiated derivative with triisopropyl borate.

The subsequent palladium-catalyzed Suzuki-Miyaura coupling between the aryl triflate **4** and the labile boronic acid **6** gave the biaryl **7** in 75% yield. Hydrolysis of the methoxymethyl ether (MOM) groups by treatment with methanolic HCl, followed by intramolecular transesterification promoted by trifluoroacetic acid, completed the synthesis of the tricyclic benzochromenone backbone and afforded the methyl ester of hapten **AL*a***, compound **8**, in 97% yield. To complete the synthesis of hapten **AL*a***, only the hydrolysis of the methyl ester moiety of **8** was required, which was initially carried out under basic conditions (LiOH in THF-H_2O at room temperature, rt). However, under these conditions, the central lactone group of the benzochromenone core was partially opened, requiring acid treatment of the reaction crude to reconstruct the tricyclic ring system. It proved more convenient to carry out this transformation using enzymatic hydrolysis, so a lipase from *Candida antarctica* immobilized on an acrylic resin was used to hydrolyze the methyl ester group, providing hapten **AL*a*** in practically quantitative yield.

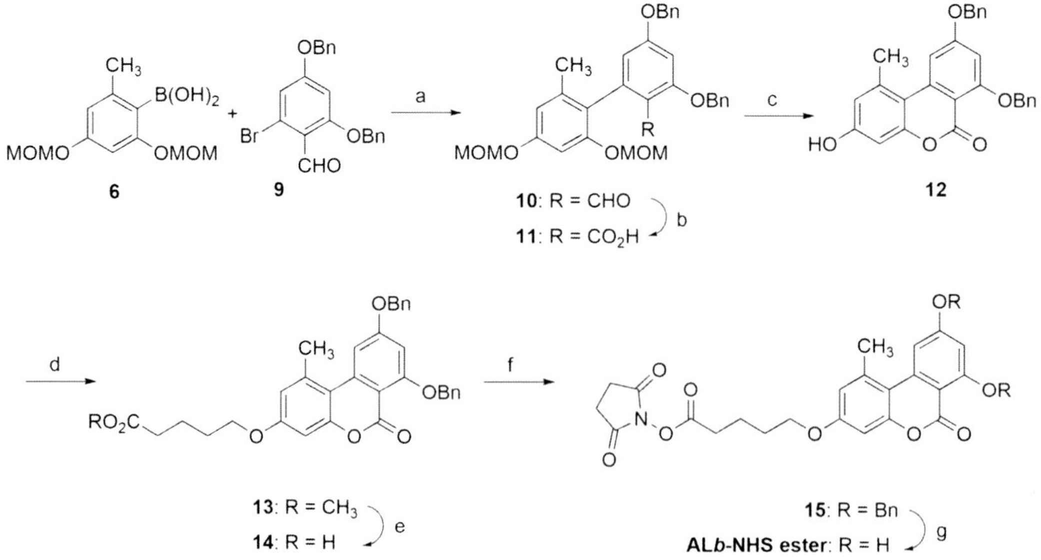

Scheme 1. Synthesis of **AL*a*-NHS** ester. Reagents and conditions: (a) $Br(CH_2)_4CO_2CH_3$, K_2CO_3, KI, Bu_4NBr, acetone, reflux, 16 h, 75% of **3**. (b) Tf_2O, pyridine, 0 °C to rt, 20 h, 91%. (c) i. n-BuLi, THF, −78 °C, 40 min; ii. $B(O^iPr)_3$, −78 °C to 0 °C, 1.5 h, 93%. (d) $Pd(PPh_3)_4$, K_2CO_3, DMF, 93 °C, 24 h, 75%. (e) i. HCl, MeOH, rt, 22 h; ii. TFA, CH_2Cl_2, rt, 20 h, 97%. (f) Lipase acrylic resin, THF-PB 100 mM, rt, 20 h, 93%. (g) EDC·HCl, NHS, DMF, rt, overnight, 99% of crude product.

Upon completion of the synthesis of hapten **AL*a***, its carboxylic group was activated by forming the corresponding *N*-hydroxysuccinimidyl ester. This transformation was carried out under conventional activation conditions, with *N*-(3-dimethylaminopropyl)-*N′*-ethylcarbodiimide hydrochloride (EDC·HCl) and *N*-hydroxisuccinimide (NHS) in *N,N*-dimethylformamide (DMF) at rt, yielding the corresponding *N*-hydroxysuccinimidyl ester, **AL*a*-NHS**, in good yield. The activated hapten was extracted essentially pure from the reaction as judged by ^1H NMR, so it was further used without additional purification by column chromatography. NMR spectra of all of the intermediates and the hapten can be found in the Supplementary Materials file.

Hapten **AL*b*** was synthesized following a similar procedure as hapten **AL*a***, except that in this case the tricyclic benzochromenone core was built first, with the hydroxyl groups appropriately protected to allow subsequent incorporation of the spacer arm at the required C-3 position. As shown in Scheme 2, the synthesis of the benzochromenone ring system began with the palladium-catalyzed cross-coupling reaction between the aryl boronic acid **6** and the previously reported bromobenzaldehyde **9** [24,25]. This coupling was carried out under conditions similar to those previously used for the conversion of **4** and **6** into **7**, obtaining the biphenyl-2-carbaldehyde **10** in 77% yield. The formyl group was further oxidized to the carboxylic group under Pinnick oxidation conditions to afford the biphenyl-2-carboxylic acid **11**, which was then treated with methanolic HCl at 55 °C to promote deprotection of the MOM groups and further intramolecular esterification, thus completing the formation of the tricyclic benzochromenone system. Under these conditions, both sequential processes worked extremely well, affording **12** in practically quantitative yield. *O*-alkylation of the phenol-like hydroxyl group at the C-3 position of **12** with methyl 5-bromovalerate, using Cs_2CO_3 in DMF as base, gave the *O*-alkylated derivative **13** in 94% yield. The methyl ester of **13** was further converted to the corresponding carboxylic group under enzymatic hydrolytic conditions, yielding **14** also in high yield. The hapten **AL*b*** was first obtained by hydrogenolysis of both benzyl ether groups of **14** using 5% Pd on activated carbon as catalyst. With hapten **AL*b*** in hand, we activated the carboxylic group using the carbodiimide-NHS procedure as was done for hapten **AL*a***. However, the overall yield from these two processes was low, most likely motivated by an intermolecular esterification reaction between a hydroxyl group and the aliphatic active ester that resulted in the spontaneous formation of a transparent thin film, a polyester polymer, on the flask walls. By reversing the order of these steps, i.e., by activating the carboxylic group first and then releasing the hydroxyl groups, a much better result was obtained. Thus, treatment of carboxylic acid **14** with EDC and NHS as before, followed by hydrogenolysis of the benzyl ether moieties of the resulting *N*-hydroxysuccinimidyl ester **15** with 5% Pd on activated carbon in acetone, gave the desired *N*-hydroxysuccinimidyl ester of hapten **AL*b***, **AL*b*-NHS** ester, with a very high overall yield. As in the case of the active ester of hapten **AL*a***, the **AL*b*-NHS** ester was extracted essentially pure from the reaction as judged by ^{1}H NMR, so it was further used without additional purification by column chromatography. NMR spectra of all of the intermediates and the hapten can be found in the Supplementary Materials file.

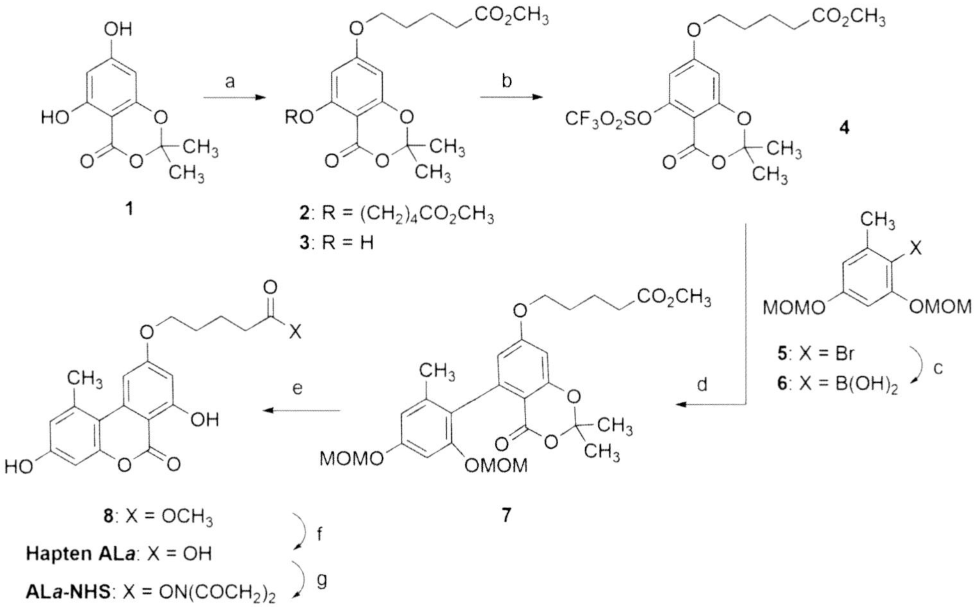

1

2: R = $(CH_2)_4CO_2CH_3$
3: R = H

4

5: X = Br
6: X = $B(OH)_2$

7

8: X = OCH_3
Hapten AL*a*: X = OH
AL*a*-NHS: X = $ON(COCH_2)_2$

Scheme 2. Synthesis of **AL*b*-NHS** ester. Reagents and conditions: (a) $Pd(PPh_3)_4$, K_2CO_3, DMF, 95 °C, 19 h, 77%. (b) $NaH_2PO_4{\cdot}H_2O$, $NaClO_2$, ^{t}BuOH-H_2O (5:1), rt, 5 h, 96%. (c) ^{i}PrOH, THF, conc HCl, 55 °C, 24 h, 98%. (d) $Br(CH_2)_4CO_2CH_3$, Cs_2CO_3, DMF, 94%. (e) Lipase acrylic resin, THF-PB 100 mM, rt, 20 h, 99%. (f) EDC·HCl, NHS, DMF, rt, overnight. (g) 5% Pd/C, acetone, H_2 (1.5 atm), rt, 19 h, 95% of crude product from **14**.

2.2. Bioconjugate Preparation

Bioconjugates of haptens **AL*a*** and **AL*b*** were prepared by the active ester method. The activated haptens were dissolved in dimethyl sulfoxide (DMSO) instead of DMF to improve the solubility. Moreover, the number of hapten equivalents required for efficiently labelling the studied proteins was higher than usual. Commonly, 20-fold hapten-to-protein molar excess for bovine serum albumin (BSA), and 10-fold excess for ovalbumin (OVA) and horseradish peroxidase (HRP) are usually employed in our laboratory. For these haptens, 40-fold and 15-fold excess was used for BSA and HRP conjugates, respectively. Moreover, extremely slow addition of the hapten over the protein solution was required. These concentrations and procedures were necessary, probably due to low hapten solubility in buffer and potential intermolecular polymerization reactions that inactivate the hapten. The obtained bioconjugates were purified by size-exclusion chromatography and characterized by matrix-assisted laser desorption ionization time-of-flight mass spectrometry (MALDI-TOF/MS) analysis. The two BSA conjugates had similar hapten densities, with hapten-to-protein molar ratios of 15.2 and 18.6 for BSA-**AL*a*** and BSA-**AL*b***, respectively, which is considered optimal for immunogens—excessive molar ratios could lead to low protein solubility, and higher or lower hapten densities could be counter-productive for high-affinity antibody generation. Regarding ovalbumin (OVA) conjugates, molar ratios were lower than those of BSA conjugates—around 3 for both haptens –, as it is desirable for coating conjugates to enhance the competitive reaction with the target analyte. Finally, the hapten densities of the enzyme tracers were estimated to be 2.0 and 2.2 for haptens **AL*a*** and **AL*b***, respectively, which is within the expected range for HRP conjugates. The MALDI spectra of the prepared bioconjugates can be seen in Figure 2.

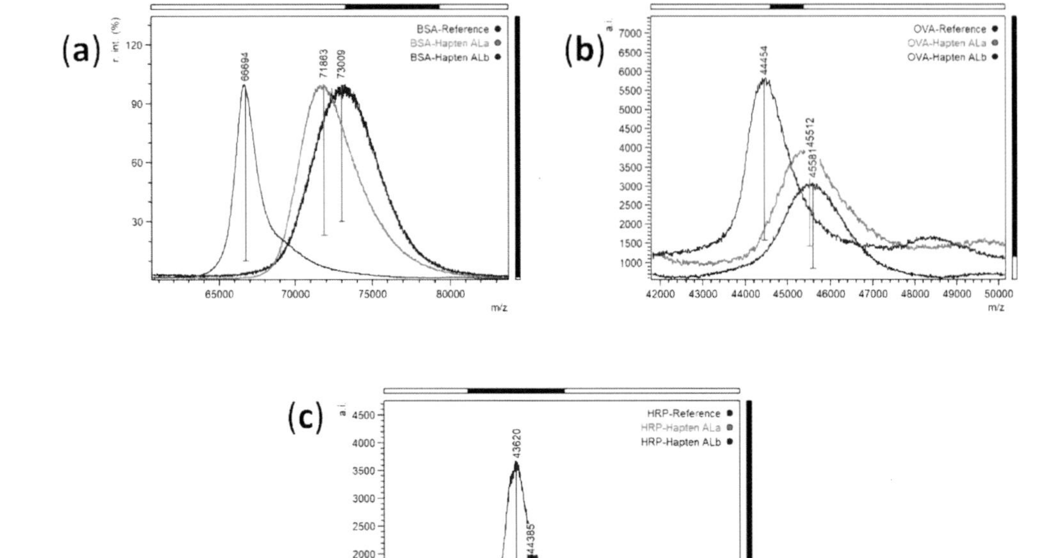

Figure 2. MALDI-TOF mass spectra (singly charged ions) of proteins (blue) and bioconjugates with hapten **AL*a*** (green) and hapten **AL*b*** (brick-red). (**a**) Normalized spectra of BSA and BSA conjugates, (**b**) Spectra of OVA and OVA conjugates, and (**c**) Spectra of HRP and HRP conjugates.

2.3. Assessment of the Immune Response

Four polyclonal antibodies were generated in this study, two from each BSA-hapten conjugate. To evaluate the immune response to the prepared synthetic haptens, binding of the antibodies to the homologous conjugate—the conjugate with the same hapten that was used to generate the corresponding antibody—was studied by checkerboard competitive ELISA, using the direct and the indirect assay formats.

Concerning direct assays, the IC_{50} values for AOH of the obtained antibodies were in the low nanomolar range (Table 1). AL*a*-type antibodies showed equal or similar IC_{50} values for AOH and AME. In particular, antibody AL*a*#1 showed very high affinity—IC_{50} values were 2.2 nM—for both mycotoxins, and the cross-reactivity (CR) values of antibodies AL*a*#1 and AL*a*#2 for AME were 100% and 199%, respectively. These are the first reported polyclonal antibodies with equivalent recognition to both *Alternaria* toxins.

To date, only one monoclonal antibody with such specificity has been published [15]. In contrast, AL*b*-type antibodies bound AOH with high affinity, but their recognition for AME was negligible—CR values were below 1% (Table 1). The IC_{50} values to AOH of these specific antibodies were 1.2 nM, an affinity comparable to that of previously published polyclonal antibodies [11,13,16]. The position of the spacer arm in hapten **AL*a*** provided a closer mimic of the alkylated hydroxyl group of AME (C-9 position), whereas in hapten **AL*b*** the hydroxyl groups at C-7 and C-9 were unsubstituted, as in the molecule of AOH (Figure 1). Therefore, display of the hydroxyl group at C-9 was maximized in hapten **AL*b***, which explains the much lower affinity of AL*b*-type antibodies for AME compared to AOH.

Table 1. Antibody characterization by checkerboard direct and indirect competitive ELISA using the corresponding homologous conjugate (n = 3) [a].

	Assay Conjugate									
	Direct Assay					Indirect Assay				
pAb	[pAb] [b]	[HRP] [c]	IC_{50} [d] AOH	IC_{50} AME	CR [e] (%)	[pAb]	[OVA]	IC_{50} AOH	IC_{50} AME	CR (%)
AL*a*#1	10	10	2.20	2.20	100	100	100	6.28	3.08	204
AL*a*#2	10	10	7.61	3.83	199	10	10	36.2	8.18	442
AL*b*#1	10	10	1.19	170	0.70	100	100	6.23	402	1.55
AL*b*#2	10	10	1.19	218	0.55	300	100	27.5	403	6.82

[a] The A_{max} values were higher than 1.0. [b] Dilution factor × 10^{-3}. [c] Bioconjugate concentration in ng/mL. [d] Values are in nM. [e] Cross-reactivity values with AOH as reference.

Regarding the indirect assay format, the four antibodies bound the corresponding homologous coating conjugate. As observed with the direct format, the AL*a*-derived antibodies recognized AOH and AME, whereas the AL*b*-derived antibodies were more specific to AOH (Table 1). The IC_{50} values were consistent with previously published results for indirect competitive ELISA with polyclonal antibodies [10,11,16]. Our strategy to prepare immunizing haptens with opposite linker tethering sites clearly demonstrated that the linker position strongly determines the specificity of antibodies to these *Alternaria* mycotoxins.

2.4. Assessment of Heterologous Conjugates

Heterologous conjugates constitute a well-known strategy for improving the sensitivity of immunoassays. To further characterize the generated antibodies, competitive assays were carried out using the heterologous conjugate, i.e., assay conjugates of haptens **AL*a*** and **AL*b*** for AL*b*- and AL*a*-type antibodies, respectively. In the direct assay format, low binding to the heterologous tracer—with the linker on the opposite side of the AOH molecule compared to the immunizing conjugate—was observed (A_{max} values were below 0.6). In contrast, the change in the linker attachment site was not detrimental to hapten recognition in the indirect format, as the four antibodies bound the corresponding heterologous coating conjugate (Table 2). Reasonably, higher antibody and/or conjugate concentrations were required with the heterologous conjugates to reach sufficient signal. The obtained IC_{50} values using the heterologous coating conjugate were mostly lower than those obtained with the homologous assays. Anyway, CR values did not significantly change with heterologous conjugates.

Table 2. Antibody characterization by checkerboard indirect competitive ELISA using the corresponding heterologous coating conjugate (n = 3) [a].

pAb	[pAb] [b]	[OVA] [c]	IC_{50} [d] AOH	IC_{50} AME	CR (%)
AL*a*#1	20	100	3.83	6.06	63.2
AL*a*#2	10	100	11.2	2.80	400
AL*b*#1	20	1000	3.32	230	1.44
AL*b*#2	30	1000	13.4	168	7.98

[a] The A_{max} values were higher than 1.0. [b] Dilution factor × 10^{-3}. [c] Bioconjugate concentration in ng/mL. [d] Values are in nM.

3. Conclusions

In this study, two de novo synthesized and purified AOH haptens were comprehensively characterized by spectrometric methods, and bioconjugates with unique structure and composition were prepared for the first time. In this perspective, it is worth noting the challenges of obtaining stable enzyme tracers with high activity. This matter was most likely caused by the chemical characteristics of *Alternaria* toxins and their haptens, which could explain why no direct competitive immunoassays for these mycotoxins have been reported up to now. Once this issue was overcome, the resultant immunoreagents were thoroughly investigated utilizing both direct and indirect competitive ELISA, as well as homologous and heterologous conjugates. Remarkably, antibodies capable of binding AOH and AME with affinities in the low nanomolar range were eventually generated from both haptens. Given that the levels of these mycotoxins are not yet regulated, both specific and generic antibodies are relevant. Our findings showed that hapten **AL*a***, with the linker at the methylated hydroxyl group in AME (C-9 position), was particularly well-suited for producing antibodies that recognized similarly both toxins, whereas antibodies generated from hapten **AL*b***, with the spacer arm at the hydroxyl group in C-3 position, primarily bound AOH. In contrast to previous one-pot hapten synthesis and bioconjugation procedures, the strategy described here for producing AOH haptens with alternative linker tethering sites not only enabled high-affinity antibodies with different specificities, but it may also help to improve the sensitivity of immunoassays to *Alternaria* mycotoxins by using site heterologous haptens.

4. Materials and Methods

4.1. Reagents and Instruments

Standard AOH [3,7,9-trihydroxy-1-methyl-benzo[c]chromen-6-one, CAS registry number 641-38-3, Mw 258.23] and AME [3,7-dihydroxy-9-methoxy-1-methyl-benzo[c]chromen-6-one, CAS registry number 23452-05-3, Mw 272.25] from *Alternaria* sp. were purchased from Merck (Darmstadt, Germany). Mycotoxins were dissolved in anhydrous *N,N*-dimethylformamide (DMF), and the stock solutions were stored at −20 °C. Phosphate buffered saline (PBS) 10× solution (Fisher BioReagents BP399-20) was from Thermo Fisher Scientific (Waltham, MA, USA). Immunizing bioconjugates were prepared with BSA, fraction V, obtained from Roche Applied Science (Mannheim, Germany). OVA, HRP, Freund's adjuvants, and adult bovine serum, were acquired from Merck (Darmstadt, Germany). Polyclonal goat anti-rabbit (GAR) immunoglobulins antibody and polyclonal goat anti-rabbit immunoglobulins antibody conjugated to peroxidase (GAR-HRP) were purchased from Rockland Immunochemicals Inc. (Pottstown, PA, USA) and BioRad (Madrid, Spain), respectively. 3,3′,5,5′-Tetramethylbenzidine (TMB) liquid substrate for ELISA was obtained from Biopanda Reagents Ltd. (Belfast, UK). Other reagents, materials, and instruments employed for bioconjugate preparation and ELISA experiments are described in the Supplementary Materials file.

4.2. Synthesis of the N-hydroxysuccinimidyl Ester of Hapten **ALa**

4.2.1. Preparation of methyl 5-((5-hydroxy-2,2-dimethyl-4-oxo-4H-benzo[d][1,3]dioxin-7-yl)oxy)pentanoate (**3**)

Methyl 5-bromovalerate (186 μL, 266 mg, 1.36 mmol, 1.1 equiv) was added to a solution of 1,3-benzodioxinone **1** (260 mg, 1.237 mmol), KI (83 mg, 0.500 mmol, 0.4 equiv), Bu_4NBr (0.5 mg, 1.6 μmol) y K_2CO_3 (188 mg, 1.36 mmol, 1 equiv) in dry acetone (9 mL) under nitrogen. After heating the mixture at reflux for 16 h, the acetone was eliminated at reduced pressure and the resulting brownish residue was diluted with water and extracted with Et_2O. The combined organic layers were washed with water and brine, dried over anhydrous $MgSO_4$ and concentrated under vacuum. The obtained crude product was purified by chromatography on silica gel, using hexane-EtOAc mixtures from 9:1 to 7:3 as eluent, to afford, in order of elution, dialkylated derivative **2** (67.7 mg, 12.5%) and monoalkylated compound **3** (300 mg, 75%) as a white solid. Mp 97.3–98.2 °C (crystallized from hexane-EtOAc) IR (ATR) ν_{max} (cm^{-1}) 3017 (w), 1740 (s), 1672 (s), 1251 (s), 1159 (s), 840 (s), 794 (s); ^1H NMR (300 MHz, $CDCl_3$) δ 10.42 (s, 1H, OH), 6.11 (d, J = 2.2 Hz, 1H, H-6), 5.97 (d, J = 2.2 Hz, 1H, H-8), 3.98 (t, J = 5.8 Hz, 2H, H_2-5), 3.68 (s, 3H, OCH_3), 2.39 (t, J = 7.0 Hz, 2H, H_2-2), 1.81 (m, 4H, H_2-3 and H_2-4), 1.72 (s, 6H, 2×CH_3); ^{13}C NMR (75 MHz, $CDCl_3$) δ 173.8 (CO_2CH_3), 167.2 (CO), 165.3 (C-7), 163.2 (C-8a), 156.9 (C-5), 107.0 (C-2), 96.3 (CH-6), 95.1 (CH-8), 93.1 (C-4a), 68.1 (CH_2-5), 51.7 (OCH_3), 33.7 (CH_2-2), 28.4 (CH_2-4), 25.8 (2×CH_3), 21.6 (CH_2-3); HRMS (TOF MS ES+) m/z calculated for $C_{16}H_{20}O_7$ $[M + H]^+$ 325.1282, found 325.1283.

4.2.2. Preparation of methyl 5-((2,2-dimethyl-4-oxo-5-(((trifluoromethyl)sulfonyl)oxy)-4H-benzo[d][1,3]dioxin-7-yl)oxy)pentanoate (**4**)

Triflic anhydride (230 μL, 1.369 mmol, 1.5 equiv) was added to a solution of phenol **3** (296 mg, 0.913 mmol) in anhydrous pyridine (4.5 mL) at 0 °C under nitrogen. The reaction mixture was allowed to warm to rt and stirred for 20 h, then cooled down to 0 °C and treated with a saturated aqueous solution of $NaHCO_3$, stirred for a few minutes at rt and then extracted with Et_2O. The organic layers were washed with water, a 1% (w/v) aqueous solution of $CuSO_4$ and brine, dried over anhydrous $MgSO_4$ and concentrated at reduced pressure. The obtained residue was chromatographed on silica gel, using hexane-EtOAc mixtures from 9:1 to 8:2 as eluent, to give aryl triflate **4** (378.8 mg, 91%) as a white semisolid. IR (ATR) ν_{max} (cm^{-1}) 3114 (w), 1746 (s), 1733 (s), 1381 (s), 1228 (s), 1167 (s), 869 (s); ^1H NMR (300 MHz, $CDCl_3$) δ 6.51 (d, J = 2.3 Hz, 1H, H-6), 6.45 (d, J = 2.3 Hz, 1H, H-8), 4.02 (t, J = 5.8 Hz, 2H, H_2-5), 3.68 (s, 3H, CO_2CH_3), 2.40 (t, J = 6.9 Hz, 2H, H_2-2), 1.85 (m, 4H, H_2-3 and H_2-4), 1.73 (s, 6H, 2xCH_3); ^{13}C NMR (75 MHz, $CDCl_3$) δ 173.7 (CO_2CH_3), 165.0 (CO), 158.9 (C-7), 157.2 (C-8a), 150.1 (C-5), 106.7 (C-2), 105.7 (CH-6), 101.6 (CH-8), 101.0 (C-4a), 68.9 (CH_2-5), 51.8 (OCH_3), 33.6 (CH_2-2), 28.3 (CH_2-4), 25.7 (2×CH_3), 21.5 (CH_2-3); ^{19}F NMR (282 MHz, $CDCl_3$) δ 73.1 (s, CF_3); HRMS (TOF, ES+) m/z calculated for $C_{17}H_{23}F_3NO_9S$ $[M + NH_4]^+$ 474.1040; found 474.1027.

4.2.3. Preparation of methyl 5-((5-(2,4-bis(methoxymethoxy)-6-methylphenyl)-2,2-dimethyl-4-oxo-4H-benzo[d][1,3]dioxin-7-yl)oxy)pentanoate (**7**)

(i) Preparation of boronic acid **6**. A solution of n-BuLi in hexane (1.3 M, 336 μL, 0.436 mmol, 1.05 equiv) was dropwise added to a solution of aryl bromide **5** (122.3 mg, 0.420 mmol) in anhydrous THF (2.5 mL) at −78 °C under nitrogen. The reaction mixture was stirred at this temperature for 40 min, $B(O^iPr)_3$ (322 μL, 1.386 mmol, 3.3 equiv) was then added and the mixture stirred for 1.5 h. After this time, the dry ice bath was replaced by an ice bath and the mixture treated with an aqueous saturated solution of NH_4Cl (0.7 mL), then diluted with water and extracted with Et_2O. The organic layers were washed with brine, dried over anhydrous Na_2SO_4 and concentrated under reduced pressure to give boronic acid **6** (100.0 mg, 93%) as a thick oil that was immediately used in the next reaction without further purification since it is relatively prone to protodeboronation [26]. ^1H NMR (300 MHz, DMSO-d_6) δ 7.91 (s, 1H, BOH), 6.49 (d, J = 2.1 Hz, 1H, H-4), 6.47 (d,

J = 2.10 Hz, 1H, H-6), 5.12 and 5.09 (each s, 2H each, 2×OCH_2O), 3.37 and 3.34 (each s, 3H each, 2×OCH_3), 2.20 (s, 3H, CH_3).

(ii) Coupling reaction between aryl triflate **4** and boronic acid **6**. A mixture of the above obtained boronic acid **6** (47.4 mg, 0.185 mmol), aryl triflate **4** (41.6 mg, 0.091 mmol), powdered K_2CO_3 (43.2 mg, 0.312 mmol) and $Pd(PPh_3)_4$ (11.4 mg, 9.9 μmol) under nitrogen was dissolved in anhydrous DMF (1.2 mL), previously degassed by three freeze-vacuum-thaw cycles. The mixture was heated at 93 °C and stirred at this temperature for 24 h. The mixture was cooled to rt, quenched with water and extracted with EtOAc. The combined organic layers were successively washed with water, a 1.5% (w/v) aqueous solution of LiCl and brine, and dried over anhydrous $MgSO_4$. The obtained residue after evaporation of the solvent was chromatographed on silica gel, using hexane-EtOAc 8:2 as eluent, to afford biaryl compound **7** (35.2 mg, 75%) as a yellowish oil. ^1H NMR (300 MHz, $CDCl_3$) δ 6.71 (d, J = 2.3 Hz, 1H, H-6), 6.64 (d, J = 2.3 Hz, 1H, H-8), 6.42 (d, J = 2.5 Hz, 1H, H-5), 6.40 (d, J = 2.5 Hz, 1H, H-3), 5.18 (AB system, J = 6.7 Hz, 2H, OCH_2O), 4.98 (AB system, J = 6.6 Hz 2H, OCH_2O), 3.99 (t, J = 5.6 Hz, 2H, H_2-5), 3.67 (s, 3H, CO_2CH_3), 3.50 and 3.29 (each s, 3H each, 2×OCH_3), 2.39 (t, J = 6.8 Hz, 2H, H_2-2), 2.04 (s, 3H, CH_3 Ph), 1.81 (m, 4H, H_2-3 and H_2-4), 1.71 (s, 6H, 2×CH_3); ^{13}C NMR (75 MHz, $CDCl_3$) δ 173.9 (CO_2CH_3), 164.1 (CO), 159.2 (C-7), 158. 5 (C-8a), 157.4 (C-4), 154.8 (OC-2), 142.9 (C-6), 136.9 (C-5), 123.9 (C-1), 113.8 (CH-3), 110.6 (CH-8), 106.8 (C-2), 105.1 (CH-5), 101.2 (CH-6), 94.9 and 94.7 (2×OCH_2O), 68.1 (CH_2-5), 56.3 and 55.9 (2×OCH_3), 51.8 (CO_2CH_3), 33.7 (CH_2-2), 28.6 (CH_2-4), 26.3 and 25.2 (2×CH_3), 21.7 (CH_2-3), 20.7 (CH_3 Ph). HRMS (TOF, ES+) m/z calculated for $C_{27}H_{34}O_{10}$ $[M + H]^+$ 519.2225; found 519.2212.

4.2.4. Preparation of methyl 5-((3,7-dihydroxy-1-methyl-6-oxo-6H-benzo[c]chromen-9-yl)oxy)pentanoate (8)

A 3 M solution of HCl in MeOH (150 μL, 0.450 mmol) was added to a solution of biaryl compound **7** (26.1 mg, 0.050 mmol) in anhydrous MeOH (1.5 mL) and the reaction mixture was stirred at rt for 22 h. After concentration under vacuum, the residue was dissolved in anhydrous CH_2Cl_2 (4 mL) and treated with trifluoroacetic acid (430 μL). Following stirring for 20 h at rt, thin layer chromatography showed the formation of a single compound and all the volatiles were removed under vacuum, using $CHCl_3$ to co-evaporate the last traces of TFA. The obtained residue was purified by chromatography, using $CHCl_3$ as eluent, to give benzochromenone derivative **8** (18.1 mg, 97%) as a white solid. ^1H NMR (300 MHz, DMSO-d_6) δ 11.79 and 10.33 (each s, 1H each, 2×OH), 7.13 (d, J = 2.2 Hz, 1H, H-10), 6.69 (d, J = 2.6 Hz, 1H, H-2), 6.61 (d, J = 2.6 Hz, 1H, H-4), 6.55 (d, J = 2.2 Hz, 1H, H-8), 4.11 (t, J = 5.9 Hz, 2H, H_2-5), 3.59 (s, 3H, OCH_3), 2.68 (s, 3H, CH_3), 2.41 (t, J = 7.0 Hz, 2H, H_2-2), 1.85–1.56 (m, 4H, H_2-3 and H_2-4); ^{13}C NMR (75 MHz, DMSO-d_6) δ 173.2 (CO_2CH_3), 165.5 (CO), 164.6 (C-9), 164.1 (C-7), 158.5 (C-3), 152.6 (C-4a), 138.4 (C-1), 137.7 (C-10a), 117.5 (CH_2-2), 108.8 (C-10b), 103.6 (CH-10), 101.6 (CH-4), 99.5 (CH-8), 98.3 (C-6a), 67.8 (CH_2-5), 51.2 (CO_2CH_3), 32.8 (CH_2-2), 27.8 (CH_2-4), 25.0 (CH_3), 21.1 (C-3); HRMS (TOF, ES+) m/z calculated for $C_{20}H_{21}O_7$ $[M + H]^+$ 373.1282; found 373.1278.

4.2.5. Preparation of 5-((3,7-dihydroxy-1-methyl-6-oxo-6H-benzo[c]chromen-9-yl)oxy) pentanoic acid (Hapten AL*a*)

Lipase from *Candida antarctica* immobilized on acrylic resin (23 mg) was added to a solution of methyl ester **8** (16.6 mg, 0.0446 mmol) in a 4:1 mixture of 100 mM sodium phosphate buffer (pH 7.4) and THF (1.5 mL) at 30 °C. The resulting heterogeneous mixture was smoothly stirred for 24 h at rt and then filtered to separate the enzyme. The filtrated and washing THF phases were combined, diluted with EtOAc, washed with brine, dried over anhydrous $MgSO_4$, and concentrated *in vacuo* to afford hapten **AL*a*** (14.9 mg, 93%) as a white amorphous solid. ^1H NMR (300 MHz, THF-d_8) δ 11.99 and 9.19 (each s, 1H each, 2×OH), 7.27 (d, J = 2.2 Hz, 1H, H-10), 6.67 (d, J = 2.7 Hz, 1H, H-2), 6.61 (d, J = 2.6 Hz, 1H, H-4), 6.56 (d, J = 2.2 Hz, 1H, H-8), 4.13 (t, J = 6.1 Hz, 2H, H_2-5), 2.78 (s, 3H, CH_3), 2.33 (t, J = 7.1 Hz, 2H, H_2-2), 1.91–1.77 (m, 4H, H_2-3 and H_2-4); ^{13}C NMR (126 MHz, THF-d_8) δ 174.5 (CO_2H), 167.1 (CO), 166.4 (C-9), 166.2 (C-7), 159.9 (C-3), 154.5 (C-4a), 139.5 (C-1), 139.2

(C-10a), 118.5 (CH-2), 110.7 (C-10b), 105.1 (CH-10), 102.8 (CH-4), 100.3 (CH-8), 100.0 (C-6a), 69.2 (CH_2-5), 34.0 (CH_2-2), 29.6 (CH_2-4), 25.0 (CH_3, overlapped with solvent signal), 22.6 (CH_2-3); HRMS (TOF, ES+) *m/z* calculated for $C_{19}H_{18}O_7$ $[M + H]^+$ 359.1125; found 359.1122.

4.2.6. Preparation of 2,5-dioxopyrrolidin-1-yl 5-((3,7-dihydroxy-1-methyl-6-oxo-6H-benzo [c]chromen-9-yl)oxy)pentanoate (**AL*a*-NHS** Ester)

A solution of hapten **AL*a*** (11.0 mg, 30.7 μmol), *N*-(3-dimethylaminopropyl)-*N'*-ethylcarbodiimide hydrochloride (EDC·HCl) (7.0 mg, 36.8 μmol, 1.2 equiv) and *N*-hydroxisuccinimide (NHS) (5.0 mg, 43.4 μmol, 1.4 equiv) in anhydrous DMF (0.6 mL) was stirred at rt under nitrogen overnight. The reaction mixture was diluted with CH_2Cl_2, washed with water, a 1.5% (*w*/*v*) aqueous solution of LiCl and brine, dried over anhydrous $MgSO_4$ and concentrated under reduced pressure to give the *N*-hydroxysuccinimidyl ester of hapten **AL*a***, **AL*a*-NHS** ester, (13.8 mg, ca. 99% of crude product) as a slightly yellowish oil which was used immediately for the preparation of the corresponding protein bioconjugates. ^{1}H NMR (500 MHz, THF-d_8) δ 11.99 and 9.06 (each s, 1H each, 2×OH), 7.28 (d, *J* = 2.2 Hz, 1H, H-10), 6.66 (d, *J* = 2.6 Hz, 1H, H-2), 6.60 (d, *J* = 2.7 Hz, 1H, H-4), 6.57 (d, *J* = 2.2 Hz, 1H, H-8), 4.17 (t, *J* = 5.8 Hz, 2H, H_2-5), 2.78 (s, 3H, CH_3), 2.75 (br s, 4H, $COCH_2CH_2CO$), 2.72 (t, *J* = 7.0 Hz, 2H, H_2-2), 1.95 (m, 4H, H_2-3 and H_2-4).

4.3. *Synthesis of the N-hydroxysuccinimidyl Ester of Hapten* ***ALb***

4.3.1. Preparation of 3,5-bis(benzyloxy)-2',4'-bis(methoxymethoxy)-6'-methyl-[1,1'-biphenyl]-2-carbaldehyde (**10**)

An ampoule containing a mixture of freshly prepared aryl boronic acid **6** (104.5 mg, 0.408 mmol, 2 equiv), 2,4-*bis*(benzyloxy)-6-bromobenzaldehyde **9** (80.9 mg, 0.204 mmol), K_2CO_3 (63.6 mg, 0.460 mmol, 2.2 equiv) and $Pd(PPh_3)_4$ (26.6 mg, 0.023 mmol, 0.1 equiv) in anhydrous DMF (2 mL) was exhaustively degassed by freeze-thaw cycles. The ampoule was closed under vacuum and heated at 95 °C for 19 h. After cooling, the ampoule was opened and the reaction mixture was poured onto water and extracted with EtOAc. The combined organic extracts were washed with water, a 1.5% (*w*/*v*) aqueous solution of LiCl and brine, dried under anhydrous $MgSO_4$ and concentrated under vacuum. The resulting crude reaction mixture was chromatographed on silica gel to give biaryl-2-carbaldehyde **10** (93.6 mg, 77%) as a viscous yellowish oil. ^{1}H NMR (500 MHz, $CDCl_3$) δ 10.02 (s, 1H, CHO), 7.54–7.48 (m, 2H, 2×CH Ph), 7.44–7.36 (m, 6H, 6xCH Ph), 7.36–7.29 (m, 2H, 2×CH Ph), 6.72 (d, *J* = 2.4 Hz, 1H, H-6), 6.65 (d, *J* = 2.3 Hz, 1H, H-4), 6.64 (d, *J* = 2.3 Hz, 1H, H-5'), 6.37 (d, *J* = 2.3 Hz, 1H, H-3'), 5.21–5.16 (m, two overlapped AB systems, 4H, OCH_2O and OCH_2Ph), 5.09 and 5.06 (AB system, *J* = 11.7 Hz, 1H each, OCH_2Ph), 5.08 and 4.97 (AB system, *J* = 6.7 Hz, 1H each, OCH_2O), 3.51 and 3.27 (each s, 3H each, 2×OCH_3), 1.96 (s, 3H, CH_3 Ph); ^{13}C NMR (126 MHz, $CDCl_3$) δ 189.6 (CHO), 163.5 (C-3), 162.1 (C-5), 157.6 (C-4'), 155.1 (C-2'), 144.8 (C-6'), 138.1 (C-1), 136.4 and 136.1 (2×C Ph), 128.9 (2×CH Ph), 128.8 (2×CH Ph), 128.4 (CH Ph), 128.1 CH Ph), 127.7 (2×CH Ph), 127.2 (2×CH Ph), 123.1 (C-1'), 118.6 (C-2), 110.7 (CH-5'), 109.6 (CH-3'), 101.2 (CH-6), 100.1 (CH-4), 94.7 and 94.6 (2×OCH_2O), 70.7 and 70.4 (2×OCH_2Ph), 56.3 and 56.1 (2×OCH_3), 20.6 (CH_3 Ph); HRMS (TOF, ES+) *m/z* calculated for $C_{32}H_{33}O_7$ $[M + H]^+$ 529.2221, found 529.2205.

4.3.2. Preparation of 3,5-bis(benzyloxy)-2',4'-bis(methoxymethoxy)-6'-methyl-[1,1'-biphenyl]-2-carboxylic Acid (**11**)

$NaH_2PO_4{\cdot}H_2O$ (58.6 mg, 0.425 mmol, 2.8 equiv), 2-methylbut-2-ene (322.1 μL, 3.04 mmol, 20 equiv) and $NaClO_2$ (45.3 mg, 0.501 mmol, 3.3 equiv) were successively added to a solution of biaryl-2-carbaldehyde **10** (80.4 mg, 0.152 mmol) in ^{t}BuOH (3.2 mL) and milli-Q water (0.4 mL) at 0 °C. The mixture was allowed to warm at rt and stirred for 5 h, then diluted with an aqueous saturated solution of NH_4Cl and extracted with EtOAc. The combined organic layers were washed with brine and dried over anhydrous $MgSO_4$. Chromatography on silica gel of the residue left after evaporation of the solvent at reduced pressure, using 8:2 hexane-EtOAc as eluent, gave the biaryl-2-carboxylic acid

11 (79.5 mg, 96%) as a semi solid. ^{1}H NMR (500 MHz, $CDCl_3$) δ 9.32 (s, 1H, CO_2H), 7.61–7.32 (m, 10H, 10xCH Ph), 6.68 (d, J = 2.4 Hz, 2H, H-6 and H-4), 6.65 (d, J = 2.4 Hz, 1H, H-5′), 6.44 (d, J = 2.3 Hz, 1H, H-3′), 5.21–5.14 (m, two overlapped AB systems, 4H, OCH_2O and OCH_2Ph), 5.07 and 5.04 (AB system, J = 11.7 Hz, 1H each, OCH_2Ph), 4.99 (br s, 2H, OCH_2O), 3.50 and 3.16 (each s, 3H each, 2×OCH_3), 2.01 (s, 3H, CH_3 Ph); ^{13}C NMR (126 MHz, $CDCl_3$) δ 165.7 (CO_2H), 161.1 (C-3), 157.8 (C-5), 157.6 (C-4′), 155.1 (C-2′), 141.3 (C-6′), 138.4 and 135.7 (2×C Ph), 128.9 (2×CH Ph), 128.8 (2×CH Ph), 128.6 (CH Ph), 128.4 (CH Ph), 127.7 (2×CH Ph), 127.6 (2×CH Ph), 125.1 (C-2), 115.9 (C-1′), 111.6 (CH-5′), 109.9 (CH-3′), 102.6 (CH-6), 100.3 (CH-4), 96.1 and 94.7 (2×OCH_2O), 71.5 and 70.4 (2×OCH_2Ph), 56.3 and 56.1 (2×OCH_3), 20.5 (CH_3 Ph); HRMS (TOF, ES+) m/z calculated for $C_{32}H_{33}O_8$ $[M + H]^+$ 545.2170, found 545.2156.

4.3.3. Preparation of 7,9-bis(benzyloxy)-3-hydroxy-1-methyl-6H-benzo[c]chromen-6-one (**12**)

A 50:1 (v/v) mixture of ^{i}PrOH and concentrated HCl (1.7 mL) was added to a solution of biaryl-2-carboxylic acid **11** (69.4 mg, 0.127 mmol) in THF (5.1 mL) at rt under nitrogen. The mixture was thermostated at 55 °C in an oil bath and stirred at this temperature for 24 h. After this time, the mixture was cooled to rt, diluted with a concentrated aqueous solution of $NaHCO_3$ and extracted with Et_2O. The organic phase was washed with brine, dried over anhydrous $MgSO_4$ and concentrated under vacuum to give 7,9-*bis*(benzyloxy)alternariol **12** (54.7 mg, 98%) as an amorphous whitish solid. The crude reaction product thus obtained was sufficiently pure, as judged by its NMR spectroscopic data, to be used in the next step without further purification. ^{1}H NMR (500 MHz, DMSO-d_6) δ 7.58 (d, J = 7.4 Hz, 2H, 2×CH Ph), 7.48 (d, J = 7.1 Hz, 2H, 2×CH Ph), 7.44–7.31 (m, 6H, 6xCH Ph), 7.28 (d, J = 2.2 Hz, 1H, H-10), 6.90 (d, J = 2.2 Hz, 1H, H-8), 6.63 (d, J = 2.7 Hz, 1H, H-2), 6.53 (d, J = 2.7 Hz, 1H, H-4), 5.31 and 5.29 (each s, 2H each, 2×OCH_2Ph), 2.63 (s, 3H, CH_3 Ph); ^{13}C NMR (126 MHz, DMSO-d_6) δ 163.6 (CO), 162.4 (C-9), 158.4 (C-7), 156.5 (C-3), 153.7 (C-4a), 140.0 (C-1), 138.0 (C-10a), 136.8 and 136.3 (2×CH Ph), 128.7 (2×CH Ph), 128.5 (2×CH Ph), 128.3 (CH Ph), 127.9 (2×CH Ph), 127.7 (CH Ph), 127.0 (2×CH Ph), 116.7 (CH-2), 109.1 (C-10b), 103.2 (C-6a), 102.8 (CH-10), 100.9 (CH-4), 99.8 (CH-8), 70.1 and 69.9 (2×OCH_2Ph), 25.0 (CH_3 Ph); HRMS (TOF, ES+) m/z calculated for $C_{28}H_{23}O_5$ $[M + H]^+$ 439.1540, found 439.1530.

4.3.4. Preparation of methyl 5-((7,9-bis(benzyloxy)-1-methyl-6-oxo-6H-benzo[c]chromen-3-yl)oxy)pentanoate (**13**)

Methyl bromovalerate (29.5 mg, ca. 22 µL, 0.151 mmol, 1.1 equiv) was added via syringe to a stirred suspension of Cs_2CO_3 (57.8 mg, 0.177 mmol, 1.3 equiv) and phenol **12** (60.1 mg, 0.137 mmol) in anhydrous DMF (2 mL) at rt under nitrogen and the mixture was stirred for 19 h. The resulting pale yellowish reaction mixture was diluted with water and extracted with EtOAc. The combined organic extracts were washed successively with water, a 1.5% (w/v) aqueous solution of LiCl and brine, dried over anhydrous $MgSO_4$ and concentrated under reduced pressure. The crude reaction product was purified by chromatography on silica gel, using $CHCl_3$ as eluent, to afford the *O*-alkylated product **13** (71.4 mg, 94%) as a pale yellowish semi-solid. ^{1}H NMR (500 MHz, $CDCl_3$) δ 7.59 (m, 2H, CH Ph), 7.43–7.34 (m, 8H, 8×CH Ph), 7.30 (d, J = 2.3 Hz, 1H, H-2), 6.67 (d, J = 2.7 Hz, 1H, H-10), 6.64 (d, J = 4.6 Hz, 2H, H-4), 6.64 (s, 1H, H-8), 5.28 and 5.16 (each s, 2H each, 2×OCH_2Ph), 4.00 (t, J = 5.5 Hz, 2H, H_2-5), 3.68 (s, 3H, CO_2CH_3), 2.67 (s, 3H, CH_3), 2.41 (t, J = 6.9 Hz, 2H, H_2-2), 1.84 (m, 4H, H_2-3 and H_2-4); ^{13}C NMR (126 MHz, $CDCl_3$) δ 173.9 (CO_2CH_3), 163.7 (CO), 162.9 (C-9), 159.5 (C-7), 157.8 (C-3), 154.3 (C-4a), 140.7 (C-1), 137.4 (C-10a), 136.5 and 135.9 (2×C Ph), 129.0 (2×CH Ph), 128.8 (2×CH Ph), 128.6 (CH Ph), 127.9 (CH Ph), 127.5 (2×CH Ph), 126.8 (2×CH Ph), 116.7 (CH-2), 111.0 (C-10b), 104.5 (C-6a), 103.9 (CH-10), 100.0 (CH-4), 99.9 (CH-8), 71.0 and 70.5 (each OCH_2Ph), 67.7 (CH_2-5), 51.7 (OCH_3), 33.8 (CH_2-2), 28.6 (CH_2-4), 25.6 (CH_3), 21.7 (CH_2-3); HRMS (TOF, ES+) m/z calculated for $C_{34}H_{33}O_7$ $[M + H]^+$ 553.2221, found 553.2112.

4.3.5. Preparation of 5-((7,9-bis(benzyloxy)-1-methyl-6-oxo-6H-benzo[c]chromen-3-yl)oxy) pentanoic Acid (**14**)

The hydrolysis of the methyl ester moiety of **13** was performed following the same procedure reported for the hydrolysis of ester **8** to obtain hapten **AL*a***. The methyl ester **13** (17.5 mg, 0.032 mmol), lipase from *Candida antarctica* immobilized on acrylic resin (16 mg) and a 4:1 mixture of 100 mM sodium phosphate buffer (pH 7.4) and THF (1.1 mL). Workup as described for the hydrolysis of **8** yielded acid **14** (17.0 mg, 99%) as a whitish semi-solid. ^{1}H NMR (500 MHz, THF-d_8) δ 7.68 (m, 2H, 2×CH Ph), 7.46 (m, 2H, 2×CH Ph), 7.40–7.30 (m, 6H, 4×CH Ph and H-2), 7.25 (br t, *J* = 7.5 Hz, 1H, CH Ph), 6.85 (d, *J* = 2.2 Hz, 1H, H-10), 6.71 (d, *J* = 2.8 Hz, 1H, H-4), 6.69 (d, *J* = 2.8 Hz, 1H, H-8), 5.26 (s, 4H, 2×OCH$_2$Ph), 4.05 (t, *J* = 6.2 Hz, 2H, H$_2$-5), 2.71 (s, 3H, Ar-CH$_3$), 2.32 (t, *J* = 7.2 Hz, 2H, H$_2$-2), 1.78 (m, 4H, H$_2$-3 and H$_2$-4). ^{13}C NMR (126 MHz, THF-d_8) δ 174.5 (CO$_2$H), 164.8 (CO), 163.9 (C-9), 160.9 (C-7), 156.7 (C-3), 155.7 (C-4a), 141.4 (C-1), 138.5 (C-10a), 138.3 and 137.9 (each C Ph), 129.5 (2×CH Ph), 129.2 (2×CH Ph), 129.0 (CH Ph), 128.5 (2×CH Ph), 128.2 (CH Ph), 127.6 (2×CH Ph), 117.2 (CH-2), 111.8 (C-10b), 105.4 (C-6a), 104.5 (CH-10), 100.7 (CH-4), 100.5 (CH-8), 71.4 and 71.1 (each OCH$_2$Ph), 68.8 (CH$_2$-5), 34.0 (CH$_2$-2), 29.7 (CH$_2$-4), 26.5 (CH$_3$), 22.6 (CH$_2$-3); HRMS (TOF, ES+) *m/z* calculated for $C_{33}H_{31}O_7$ [M + H]$^+$ 539.2064, found 539.2073.

4.3.6. Preparation of 2,5-dioxopyrrolidin-1-yl 5-((7,9-dihydroxy-1-methyl-6-oxo-6H-benzo [c]chromen-3-yl)oxy)pentanoate (**AL*b*-NHS** Ester)

The acid **14** obtained in the above step (15.1 mg, 28 µmol) was transformed into the corresponding *N*-hydroxysuccinimidyl ester **15** (17.2 mg) following the same procedure previously described for the transformation of hapten **AL*a*** into **AL*a*-NHS** ester, using EDC·HCl (6.4 mg, 33.6 µmol, 1.2 equiv) and NHS (4.2 mg, 36.5 µmol, 1.3 equiv) in anhydrous DMF (1 mL). ^{1}H NMR (300 MHz, CDCl$_3$) δ 7.59 (m, 2H, 2×CH Ph), 7.45–7.32 (m, 8H, 8xCH Ph), 7.31 (d, *J* = 2.3 Hz, 1H, H-2), 6.69 (d, *J* = 2.7 Hz, 1H, H-10), 6.67 (d, *J* = 2.7 Hz, 1H, H-4), 6.65 (d, *J* = 2.2 Hz, 1H, H-8), 5.29 and 5.17 (each s, 2H each, 2×OCH$_2$Ph), 4.04 (t, *J* = 5.6 Hz, 2H, H$_2$-5), 2.85 (br s, 4H, COCH$_2$CH$_2$CO), 2.72 (t, *J* = 6.9 Hz, 1H, H$_2$-2), 2.68 (s, 3H, CH$_3$), 1.96 (m, 4H, H$_2$-3 and H$_2$-4).

Thereafter, a suspension of 5% Pd/C (8 mg) and **15** in acetone (3 mL) was degassed and purged with hydrogen by several cycles of freeze-pump-thaw using a water aspirator pump. The hydrogen pressure was adjusted to 1.5 atm and the mixture was stirred vigorously overnight at rt. The reaction mixture was filtered through a disposable Teflon membrane filter (0.45 µm), and the filtrate and washing THF phases were combined and concentrated at reduced pressure to give the *N*-hydroxysuccinimidyl ester of hapten **AL*b***, **AL*b*-NHS** ester, (12.2 mg, 95% of crude product from **14**) as a viscous colorless oil which was used immediately for the preparation of the corresponding protein bioconjugates. ^{1}H NMR (500 MHz, THF-d_8/DMSO-d$_6$) δ 11.92 and 9.67 (each s, 1H each, 2×OH), 7.24 (d, *J* = 2.1 Hz, 1H, H-2), 6.83 (d, *J* = 2.7 Hz, 1H, H-4), 6.82 (d, *J* = 2.6 Hz, 1H, H-10), 6.36 (d, *J* = 2.1 Hz, 1H, H-8), 4.10 (t, *J* = 5.6 Hz, 1H, H$_2$-5), 2.79 (s, 3H, CH$_3$), 2.76 (br s, 4H, COCH$_2$CH$_2$CO), 2.71 (t, *J* = 6.9 Hz, 2H, H$_2$-2), 1.93 (m, 4H, H$_2$-3 and H$_2$-4).

4.4. Immunoreagent Preparation

Protein conjugates of the two haptens were obtained by the active ester method. A 50 mM solution of the activated hapten was prepared in DMSO. BSA and OVA solutions were prepared at 15 mg/mL in 50 mM carbonate buffer, pH 9.6. The activated hapten was added to the protein solution at 40-fold molar excess for BSA conjugates. Conjugation reactions to OVA were done with 8- or 11-fold excess for **AL*a*** and **AL*b***, respectively. Concerning HRP conjugates, the activated hapten solution (5 mM) was added over a 3 mg/mL enzyme solution in the same carbonate buffer to reach a hapten molar excess of 15. The activated haptens were added slowly to the protein solution (ca. 15–20 µL per hour), and the mixtures were incubated overnight at rt, protected from light, and with gentle stirring. Then, they were centrifuged for 5 min at 6700× *g*, and the conjugates were purified from the supernatant by size-exclusion chromatography using 100 mM phosphate buffer, pH 7.4, as eluent. Fractions containing BSA or OVA conjugates were pooled and diluted with elution buffer to a final concentration of 1 mg/mL. BSA conjugate solutions were passed through 0.45 μm sterile filters. BSA and OVA conjugate solutions were stored at −20 °C. HRP conjugate solutions were 1:1 diluted with PBS containing 1% BSA (*w*/*v*) and 0.02% (*w*/*v*) thimerosal and stored at 4 °C. The hapten-to-protein molar ratio of the prepared conjugates was determined by MALDI-TOF-MS and running BSA, OVA, and HRP for reference in the same plate, as previously described [27].

Animal manipulation was performed according to Spanish laws (RD1201/2005 and law 32/2007) and the European Directive 2010/63EU regarding the protection of experimental animals. Polyclonal antibodies to AOH and AME were obtained from the sera of immunized animals. Briefly, two female New Zealand white rabbits—weighing 2 kg at the beginning of the experiment—were immunized by four periodic subcutaneous injections of a 1:1 water-in-oil emulsion containing 300 μg of the BSA-hapten conjugate. The inoculum was prepared with complete Freund's adjuvant for the first injection and with incomplete Freund's adjuvant for subsequent injections. Boosts were applied with 21-day intervals. Animals were exsanguinated by intracardiac puncture 10 days after the last injection, and the blood was left overnight in the refrigerator at 4 °C for coagulation. Sera were separated from cells by centrifugation (3000× *g*, 20 min). Finally, immunoglobulins were partially purified by precipitation twice with one volume of cold saturated (3.9 M) ammonium sulphate solution. Antibodies were stored at 4 °C as precipitates.

4.5. Competitive ELISA Procedures

Immunoassays were carried out by competitive ELISA using the capture antibody-coated direct format and the conjugate-coated indirect format. After each incubation step, plates were washed three times with a 150 mM NaCl solution containing 0.05% (*v*/*v*) Tween 20. For direct assays, microplates were coated by overnight incubation at 4 °C with 100 μL per well of GAR solution (1 μg/mL) in 50 mM carbonate-bicarbonate buffer, pH 9.6. For indirect assays, microwells were coated with 100 μL per well of OVA-hapten conjugate solution in the same coating buffer, and overnight incubation at rt. The competitive reaction was performed by mixing in each well 50 μL of analyte solution in PBS with 50 μL of antibody dilution or enzyme tracer solution in PBS containing 0.05% (*v*/*v*) of Tween-20, and incubating 1 h at rt. For indirect assays, 100 μL per well of GAR-HRP diluted 1/10,000 in PBS containing 0.05% (*v*/*v*) of Tween-20 and 10% (*v*/*v*) of adult bovine serum was added. Signal was obtained using 100 μL per well of TMB as the chromogenic enzyme substrate and incubation at rt during 10 min. Finally, 100 μL of 1 M H_2SO_4 was added and the absorbance was read at 450 nm using 650 nm as reference wavelength.

Standard mycotoxin solutions were obtained by serially diluting in buffer the most concentrated standard solution, which was prepared from a concentrated stock solution in DMF. Eight-point standard curves were built using those solutions and a blank sample. SigmaPlot software, version 14.0 from Systat Software Inc. (San Jose, CA, USA), was employed to fit the experimental values to a standard four-parameter logistic equation. The half-maximal inhibition concentration (IC_{50}) and the maximum absorbance (A_{max}) values were considered in order to compare antibody performance. CR was calculated according to Formula (1):

$$CR\ (\%) = IC_{50}\ (AOH)/IC_{50}\ (AME) \times 100 \tag{1}$$

Supplementary Materials: The following are available online at https://www.mdpi.com/article/10.3390/toxins13120883/s1: General procedures, materials, and equipment; NMR spectra of all of the intermediates and the final activated haptens.

Author Contributions: Conceptualization, A.A.-F.; Data curation, A.A.-S., A.A.-F. and J.V.M.; Formal analysis, L.G.A.-M., A.A.-S. and C.A.; Funding acquisition, A.A.-S., A.A.-F. and J.V.M.; Investigation, L.G.A.-M. and C.A.; Methodology, A.A.-S., A.A.-F. and J.V.M.; Supervision, A.A.-S., C.A. and J.V.M.; Validation, L.G.A.-M. and C.A.; Writing—original draft, A.A.-S. and J.V.M.; Writing—review &

editing, A.A.-S., A.A.-F. and J.V.M. All authors have read and agreed to the published version of the manuscript.

Funding: This research was funded by the SPANISH MINISTERIO DE ECONOMÍA Y COMPETITIVIDAD, grant numbers AGL2015-64488 and RTI2018-096121, and cofinanced by EUROPEAN REGIONAL DEVELOPMENT FUNDS. Animal manipulation as well as mass, spectrometric, and proteomic analysis was performed at the SCSIE service of the University of Valencia. The proteomics laboratory is a member of *Proteored*.

Institutional Review Board Statement: This study was conducted according to the guidelines of the Declaration of Helsinki, and approved by the Institutional Review Board of GENERALITAT VALENCIANA (protocol code 2019/VSC/PEA/0179) on 20 August 2019.

Informed Consent Statement: Not applicable.

Data Availability Statement: Not applicable.

Acknowledgments: The technical assistance by Paula Peña-Murgui and José V. Gimeno is greatly appreciated.

Conflicts of Interest: The authors declare no conflict of interest.

References

1. Escrivá, L.; Oueslati, S.; Font, G.; Manyes, L. *Alternaria* mycotoxins in food and feed: An overview. *J. Food Qual.* **2017**, *2017*, 1569748. [CrossRef]
2. EFSA Panel on Contaminants in the Food Chain (CONTAM). Scientific Opinion on the risks for animal and public health related to the presence of *Alternaria* toxins in feed and food. *EFSA J.* **2011**, *9*, 2407. [CrossRef]
3. Arcella, D.; Eskola, M.; Gómez-Ruiz, J.A. EFSA Scientific Report. Dietary exposure assessment to *Alternaria* toxins in the European population. *EFSA J.* **2016**, *14*, 4654. [CrossRef]
4. Ostry, V. *Alternaria* mycotoxins: An overview of chemical characterization, producers, toxicity, analysis and occurrence in foodstuffs. *World Mycotoxin J.* **2008**, *1*, 175–188. [CrossRef]
5. Saha, D.; Fetzner, R.; Burkhardt, B.; Podlech, J.; Metzler, M.; Dang, H.; Lawrence, C.; Fischer, R. Identification of a polyketide synthase required for alternariol (AOH) and alternariol-9-methyl ether (AME) formation in *Alternaria alternata*. *PLoS ONE* **2012**, *7*, e40564. [CrossRef] [PubMed]
6. Solfrizzo, M. Recent advances on *Alternaria* mycotoxins. *Curr. Opin. Food Sci.* **2017**, *17*, 57–61. [CrossRef]
7. Hickert, S.; Bergmann, M.; Ersen, S.; Cramer, B.; Humpf, H.-U. Survey of *Alternaria* toxin contamination in food from the German market, using a rapid HPLC-MS/MS approach. *Mycotoxin Res.* **2016**, *32*, 7–18. [CrossRef] [PubMed]
8. Man, Y.; Liang, G.; Li, A.; Pan, L. Analytical methods for the determination of *Alternaria* mycotoxins. *Chromatographia* **2017**, *80*, 9–22. [CrossRef]
9. Chen, A.; Mao, X.; Sun, Q.; Wei, Z.; Li, J.; You, Y.; Zhao, J.; Jiang, G.; Wu, Y.; Wang, L.; et al. *Alternaria* mycotoxins: An overview of toxicity, metabolism, and analysis in food. *J. Agric. Food Chem.* **2021**, *69*, 7817–7830. [CrossRef]
10. Burkin, A.A.; Kononenko, G.P. Enzyme immunassay of alternariol for the assessment of risk of agricultural products contamination. *Appl. Biochem. Microbiol.* **2011**, *47*, 72–76. [CrossRef]
11. Ackermann, Y.; Curtui, V.; Dietrich, R.; Gross, M.; Latif, H.; Märtlbauer, E.; Usleber, E. Widespread occurrence of low levels of alternariol in apple and tomato products, as determined by comparative immunochemical assessment using monoclonal and polyclonal antibodies. *J. Agric. Food Chem.* **2011**, *59*, 6360–6368. [CrossRef]
12. Kong, D.; Xie, Z.; Liu, L.; Song, S.; Zheng, Q.; Kuang, H. Development of an immunochromatographic assay for the detection of alternariol in cereal and fruit juice samples. *Food Agric. Immunol.* **2017**, *28*, 1082–1093. [CrossRef]
13. Singh, G.; Velasquez, L.; Brady, B.; Koerner, T.; Huet, A.-C.; Delahaut, P. Development of an indirect competitive ELISA for analysis of alternariol in bread and bran samples. *Food Anal. Methods* **2018**, *11*, 1444–1450. [CrossRef]
14. Man, Y.; Liang, G.; Jia, F.; Li, A.; Fu, H.; Wang, M.; Pan, L. Development of an immunochromatographic strip test for the rapid detection of alternariol monomethyl ether in fruit. *Toxins* **2017**, *9*, 152. [CrossRef] [PubMed]
15. Wang, J.; Peng, T.; Zhang, X.; Yao, K.; Ke, Y.; Shao, B.; Wang, Z.; Shen, J.; Jiang, H. A novel hapten and monoclonal antibody-based indirect competitive ELISA for simultaneous analysis of alternariol and alternariol monomethyl ether in wheat. *Food Control* **2018**, *94*, 65–70. [CrossRef]
16. Yao, C.-Y.; Xu, Z.-L.; Wang, H.; Zhu, F.; Luo, L.; Yang, J.-Y.; Sun, Y.-M.; Lei, H.-T.; Tian, Y.-X.; Shen, Y.-D. High affinity antibody based on a rationally designed hapten and development of a chemiluminescence enzyme immunoassay for quantification of alternariol in fruit juice, maize and flour. *Food Chem.* **2019**, *283*, 359–366. [CrossRef]
17. Bräse, S.; Gläser, F.; Kramer, C.; Lindner, S.; Linsenmeier, A.M.; Masters, K.-S.; Meister, A.C.; Ruff, B.M.; Zhong, S. *The Chemistry of Mycotoxins*; Springer: Berlin/Heidelberg, Germany, 2013; Chapter 11; Volume 97, pp. 127–137.
18. Koch, K.; Podlech, J.; Pfeiffer, E.; Metzler, M. Total synthesis of alternariol. *J. Org. Chem.* **2005**, *70*, 3275–3276. [CrossRef]

19. Liang, D.; Luo, H.; Liu, Y.-F.; Hao, Z.-Y.; Wang, Y.; Zhang, C.-L.; Zhang, Q.J.; Chen, R.-Y.; Yu, D.-Q. Lysilactones A-C, three 6H-dibenzo[b,d]pyran-6-one glycosides from *Lysimachia clethroides*, total synthesis of lysilactone A. *Tetrahedron* **2013**, *69*, 2093–2097. [CrossRef]
20. Won, M.; Kwon, S.; Kim, T.-H. An efficient synthesis of alternariol. *J. Korean Chem. Soc.* **2015**, *59*, 471–474. [CrossRef]
21. Martinez-Solorio, D.; Belmore, K.A.; Jennings, M.P. Synthesis of the purported ent-pochonin J structure featuring a stereoselective oxocarbenium allylation. *J. Org. Chem.* **2011**, *76*, 3898–3908. [CrossRef]
22. Mallampudi, N.A.; Choudhury, U.M.; Mohapatra, D.K. Total synthesis of (−)-citreoisocoumarin, (−)-citreoisocoumarinol, (−)-12-epi-citreoisocoumarinol, and (−)-mucorisocoumarins A and B using a gold(I)-catalyzed cyclization strategy. *J. Org. Chem.* **2020**, *85*, 4122–4129. [CrossRef] [PubMed]
23. Hinman, A.W.; Davis, D.; Kheifets, V. Preparation of Resorcinol Compounds for Dermatological Use. U.S. Patent 20140256830 A1, 6 March 2014.
24. Marcyk, P.; Brown, L.E.; Huang, D.; Cowen, L.E.; Whitesell, L. Preparation of Resorcylate Aminopyrazole Compounds as Hsp90 Inhibitors and Uses. Thereof. Patent WO 2020/227368 A1, 6 May 2020.
25. Roedel, T.; Gerlach, H. Enantioselective synthesis of the polyketide antibiotic (3R,4S)-(−)-citrinin. *Liebigs Ann.* **1995**, *5*, 885–888. [CrossRef]
26. Lozada, J.; Liu, Z.; Perrin, D.M. Base-promoted protodeboronation of 2,6-disubstituted arylboronic acids. *J. Org. Chem.* **2014**, *79*, 5365–5368. [CrossRef] [PubMed]
27. López-Puertollano, D.; Mercader, J.V.; Agulló, C.; Abad-Somovilla, A.; Abad-Fuentes, A. Novel haptens and monoclonal antibodies with subnanomolar affinity for a classical analytical target, ochratoxin A. *Sci. Rep.* **2018**, *8*, 9761. [CrossRef] [PubMed]

MDPI
St. Alban-Anlage 66
4052 Basel
Switzerland
Tel. +41 61 683 77 34
Fax +41 61 302 89 18
www.mdpi.com

Toxins Editorial Office
E-mail: toxins@mdpi.com
www.mdpi.com/journal/toxins

Printed by BoD™ in Norderstedt, Germany